W9-AWP-015

DICTIONARY OF THEORETICAL CONCEPTS IN BIOLOGY

Keith E. Roe &
Richard G. Frederick

The Scarecrow Press, Inc.
Metuchen, N.J., & London
1981

Library of Congress Cataloging in Publication Data

Roe, Keith E
Dictionary of theoretical concepts in biology.

1. Biology—Dictionaries. I. Frederick, Richard G., joint author. II. Title.
QH302.5.R63 574'.03'21 80-19889
ISBN 0-8108-1353-X

Manufactured in the United States of America

For Eunice and Jan

Acknowledgments

We gratefully acknowledge the use of library collections of the Pennsylvania State University and thank the many people who have built them over the years. We also thank Alan Knutson for his help in searching for entries and references.

Contents

"Scientists are perennially aware that it is best not to trust theory until it is confirmed by evidence. It is equally true, as Eddington pointed out, that it is best not to put too much faith in facts until they have been confirmed by theory."—Robert MacArthur, Challenging Biological Problems.

Introduction

Biologists have traditionally given symbolic names to the theoretical concepts they work with, names that act as a form of shorthand reference for what are often complex ideas. These symbols become an important part of scientific communication, sometimes bordering on a language itself. They serve much the same purpose, abbreviation, as do the scientific epithets assigned to species that are used in lieu of lengthy pre-Linnaean diagnoses. Concept names may be descriptive or colorful, such as *competitive exclusion* or "*big mother*" *hypothesis*, or they may be eponyms that honor the original formulator of the concept, such as *Allen's rule* and *Wallace's line*.

Each theoretical concept has a history of its own in the literature—initial statement, criticism, and review. The more familiar biological concepts are known as much by the people who originated or revised them, their dates of appearance in the literature, and bibliographic citations as they are by definitions. An awareness of the named concepts in one's field is essential both for understanding its history and for communication with colleagues.

Though named concepts are of everyday use in scientific papers, references to them are often lacking, and there is no source that leads a reader into the literature surrounding each of them. Biologists must either be sufficiently familiar with the theories, laws, and rules in their fields in order to know where to look for background papers on them, or they must dig through various texts hoping to find a lead to the subject. Nonspecialists in a subject area may be at a loss to know either the application of a conceptual term or where to find earlier references to it, particularly when encountering the term out of context or when used with a presumption of familiarity. To the nonevolutionist, the *Red Queen hypothesis* may suggest a chess move or a character from Lewis Carroll (indeed, his queen is the inspiration for this name), while *Primordial Soup* may be interpreted as the first product of Campbell's.

The present work attempts to provide access to the literature on named theoretical concepts by citing original sources and reviews in which these concepts are elucidated. The reader is thus led not only to the foundation of each concept, but to its building blocks of criticism and historical analysis. With references through which to enter the literature, the reader can proceed by citation "networking" to work backward or forward in time, thus elaborating the search for relevant materials as much as desired. We have tried to exclude "factual" concepts, the definable, noncontroversial sort, which fall into the realm of terminology. However, the boundary between factual and theoretical concepts is not precise; consequently any list of these will contain examples that intergrade into one type or the other.

The 1,166 named concepts included in this work have been gleaned from journal articles, monographs, reviews, and histories of biological disciplines, primarily in the English language, published through the year 1979. Both plant and animal biology are covered, but neither human nor behavioral biology is heavily represented as these could take up a book of their own. Theoretical subject areas produce many more concepts than do experimental or descriptive fields, of course (evolution versus physiology, for example). With few exceptions, philosophical concepts, mathematical models, and experimental techniques are excluded. Current usage is emphasized, especially that of the past decade. However, some classical concepts from the nineteenth century and earlier still remain the bases for modern discussions; these have been included.

It is common for a single concept to be known by several names

"I have steadily endeavored to keep my mind free so as to give up any hypothesis, however much beloved (and I cannot resist forming one on every subject), as soon as facts are shown to be opposed to it."—
Charles Darwin, The Autobiography of Charles Darwin.

"Forming hypothesis is one of the most precious faculties of the human mind and is necessary for the development of sciences. Sometimes, however, hypotheses grow like weeds and lead to confusion instead of clarification. Then one has to clear the field, so that the operational concepts can grow and function."—Max Kleiber, The Fire of Life

(see, for example, *Primordial Soup, Gene Conversion,* or the *Neutral Theory*). We have used extensive cross-references to bring such equivalent names together; likewise those so closely related as to nearly always share discussion in any given paper; the *Red Queen hypothesis* and the *Law of Constant Extinction* are examples. To save space we have listed the set of references applying to these concepts under whichever name is most commonly cited, regardless of conceptual "levels." For instance, even though the Red Queen ranks below and is part of Constant Extinction, the former becomes the entry in this book because it is more commonly cited, probably because of its distinctive name. Specialized forms of certain concepts that are rarely encountered in the literature can often be found in reviews listed under the general term for the concept.

Some concepts have very definite origins. Many are named in the first articles in which they are discussed. Others may be discussed at some length without being named (in papers by Smith and Jones, let us say), while a subsequent paper may further elucidate the concept and bestow a descriptive name or the original author's surname(s) upon it (e.g., *Smith-Jones hypothesis*). In such clear-cut cases we have cited both the original articles and the one naming the concept. Very often, however, the origin of a concept may be shrouded in numerous premature glimmers published long before someone clearly states the generality. In such cases we have attempted to cite, when known, the first source to actually name the concept. From this paper and later reviews one can gain an understanding of the concept's history and prehistory. Some very general concepts have ill-defined beginnings or lack the thorough historical reviews that might illuminate their roots. For these we have listed available review articles rather than attempting to pinpoint origins.

Definitions are not included because they are generally inadequate, even misleading, for the understanding of theoretical concepts. Unlike the case with simple terms, for which concise definitions will suffice, one must usually read about the development, applications, and recent evolution of a conceptual idea or generality to grasp fully its significance to one's own research. The meanings and applications of concepts often change with the passage of time or from author to author. Therefore we have taken an objective approach in this book, indicating the general field in which a concept occurs and providing citations to authoritative discussions in the literature rather than trying to paraphrase these accounts with fixed definitions. In this regard W.B. Yapp has cogently written (*Nature* 167:160 [1951]): "The advice 'Define your terms' is bad counsel to give to biologists, for where the observations to be described have no sharp limits, if definition makes the argument simpler it does so at the expense of truth."

Citations are arranged chronologically under each entry to show the concept's historical development. However, all citations by a given author are listed together within an entry. Page numbers or chapters are given for books except: (1) in the case of an early reference that may form the basis for a concept but wherein the concept is never explicitly named; or (2) where the entire book or substantial part of it deals with the subject. Under each entry are listed any synonyms or concepts so closely related they are usually discussed together in papers and monographs. Such names that are capitalized can be found as entries themselves, while names in lower case are covered only with the primary entry under which they are assigned. Names of similar or conflicting concepts are listed below the citations for an entry as "confer" references. Journal title abbreviations generally follow those of the BIOSIS list, the *World List of Scientific Periodicals*, the *Botanico-Periodicum-Huntianum*, and the Royal Society's *Catalogue of Scientific Papers*.

We have attempted to treat as independent entries the clear-cut examples of homonyms but have not divided mere different applications of a single concept; for example, those applying to both plants and animals. Often the generalizations apply equally to both subjects, though each may be known by a very different set of citations in the literature.

In selecting references we have tried to include those that discuss the concept in question with some degree of thoroughness and that are in books and journals likely to be found in most academic libraries. It is not our intention, however, to fix as "correct" the names of concepts or to establish particular references as the only ones important to any concept. These will change as authors revise the printed record of knowledge and speculation. New reviews are likely to cite the primary sources given in this book, however, so that citation searching will continue to make it a usable work.

We realize the potential for unintentional omissions or errors in a compilation such as this and will be grateful to users of the book for their suggestions of possible entries and corrections for future editions.

". . . there is nothing more permanent
than a theory, and there is nothing more
temporary than a fact."—Joseph S. Fruton

List of Journals (Alphabetized by abbreviation)

Abh. Abt. Phytomorphogenese
Abhandlungen aus der Abteilung für Phytomorphogenese der Timiriaff-Instituts für Biologie (Russia)

Abh. Königl. Akad. Wiss. Berlin
Abhandlungen der Koniglichen Akademie der Wissenschaft in Berlin (Germany)

Adv. Genet.
Advances in Genetics (US)

Adv. Immunol.
Advances in Immunology (US)

Adv. Insect Physiol.
Advances in Insect Physiology (England)

Advmt. Sci.
Advancement of Science (England)

Aliso
Aliso (US)

Amer. Assoc. Adv. Sci. Publ.
American Association for the Advancement of Science Publication (US)

Amer. Biol. Teach.
American Biology Teacher (US)

Amer. J. Anat.
American Journal of Anatomy (US)

Amer. J. Bot.
American Journal of Botany (US)

Amer. J. Hum. Genet.
American Journal of Human Genetics (US)

Amer. J. Orthod.
American Journal of Orthodontics (US)

Amer. J. Sci.
American Journal of Science (US)

Amer. Midl. Nat.
American Midland Naturalist (US)

Amer. Mus. Novit.
American Museum Novitates (US)

Amer. Nat.
American Naturalist (US)

Amer. Orchid Soc. Bull.
American Orchid Society Bulletin (US)

Amer. Sci.
American Scientist (US)

Amer. Zool.
American Zoologist (US)

Anat. Anz.
Anatomischer Anzeiger (Germany)

Anat. Rec.
Anatomical Record (US)

Anim. Behav.
Animal Behaviour (England)

Ann. Acad. Sci. Colon.
Annales Académie des Sciences Coloniales (France)

Ann. Assoc. Amer. Geogr.
Annals of the Association of American Geographers (US)

Ann. Biol. Anim. Biochim. Biophys.
Annales de Biologie Animale Biochimie Biophysique (France)

Ann. Bot.
Annals of Botany (England)

Ann. Chim.
Annales de Chimie (France)

Ann. Entomol. Soc. Amer.
Annals of the Entomological Society of America (US)

Ann. Entomol. Fenn.
Annales Entomologici Fennici (Finland)

Ann. Eugen.
Annals of Eugenics (England)

Ann. Génét.
Annales de Génétique (France)

Ann. Hum. Genet.
Annals of Human Genetics (England)

Ann. Mag. Nat. Hist.
Annals and Magazine of Natural History (England)

Ann. Math. Statis.
Annals of Mathematical Statistics (US)

Ann. N.Y. Acad. Sci.
Annals of the New York Academy of Sciences (US)

Ann. Sci.
Annals of Science (England)

Ann. Sci. Nat. Bot.
Annales des Sciences Naturelles (A)
Botanique (France)

Ann. Sci. Nat. Zool.
Annales des Sciences Naturelles (B)
Zoologie (France)

Annu. Rep. Natl. Inst. Genet.
Annual Report of the National Institute of Genetics (Japan)

Annu. Rev. Biochem.
Annual Review of Biochemistry (US)

Annu. Rev. Ecol. Syst.
Annual Review of Ecology and Systematics (US)

Annu. Rev. Entomol.
Annual Review of Entomology (US)

Annu. Rev. Genet.
Annual Review of Genetics (US)

Annu. Rev. Microbiol.
Annual Review of Microbiology (US)

Annu. Rev. Phytopath.
Annual Review of Phytopathology (US)

Annu. Rev. Plant Physiol.
Annual Review of Plant Physiology (US)

Arb. Biol. Reichanst. Landw. Forstw. Berlin
Arbeiten aus der Biologischen Reichanstalt für Land- und Forstwirtschaft Berlin (Germany)

Arb. Bot. Inst. Wurz.
Arbeiten des Botanischen Instituts in Wurzburg (Germany)

Arch. Biochem. Biophys.
Archives of Biochemistry and Biophysics (US)

Arch. Biol.
Archives de Biologie (Belgium)

Arch. Entw. Mech.
Archiv für Entwicklungsmechanik der Organismen (Germany)

Arch. Int. Hist. Sci.
Archives Internationales d'Histoire des Sciences (France)

Arch. Int. Physiol. Biochim.
Archives Internationales de Physiologie et de Biochimie (Belgium)

Arch. Mikr. Anat.
Archiv für Mikroskopische Anatomie und Entwicklungsmechanik (Germany)

Arch. Neerl. Zool.
Archives Neerlandaises de Zoologie (Netherlands)

Arch. Path. Anat.
Archiv für Pathologische Anatomie und Physiologie (Germany)

Arch. Protistenk.
Archiv für Protistenkunde (Germany)

Arch. Sci. Phys. Nat.
Archives des Sciences Physiques et Naturelles, Supplément à la Bibliothèque Universelle (Switzerland)

Arch. Zool. Exp. Gén.
Archives de Zoologie Expérimentale et Générale (France)

Ark. Zool.
Arkiv för Zoologi (Sweden)

Auk
Auk (US)

Aust. J. Biol. Sci.
Australian Journal of Biological Sciences (Australia)

Aust. J. Sci.
Australian Journal of Science (Australia)

Aust. J. Sci. Res. B
Australian Journal of Scientific Research Series B Biological Sciences (Australia)

Aust. J. Zool.
Australian Journal of Zoology (Australia)

Behav. Biol.
Behavioral Biology (US)

Behaviour
Behaviour (Netherlands)

Ber. Deut. Bot. Ges.
Berichte der Deutschen Botanischen Gesellschaft (Germany)

Ber. Deut. Chem. Ges.
Berichte der Deutschen Chemische Gesellschaft (Germany)

Bibliogr. Genet.
Bibliographia Genetica (Netherlands)

Bijdr. Dierk.
Bijdragen tot de Dierkunde (Netherlands)

Biochem. Biophys. Res. Comm.
Biochemical and Biophysical Research Communications (US)

Biochem. Genet.
Biochemical Genetics (US)

Biochem. J.
Biochemical Journal (England)

Biochem. Soc. Symp.
Biochemical Society Symposia (England)

Biochem. Z.
Biochemische Zeitschrift (Germany)

Biogr. Mem. Natl. Acad. Sci.
Biographical Memoirs National Academy of Sciences (US)

Biol. Bull.
Biological Bulletin (US)

Biol. J. Linn. Soc.
Biological Journal of the Linnean Society (England)

Biol. Meddr.
Biologiske Meddelelser (Denmark)

Biol. Rev.
Biological Reviews (England)

Biol. Symp.
Biological Symposia (US)

Biol. Zblt.
Biologisches Zentralblatt (Germany)

Biol. Zhur.
Biologicheskii Zhurnal (Russia)

Biologist
Biologist (US)

Biometrika
Biometrika (England)

Biophys. J.
Biophysical Journal (US)

Bioscience
Bioscience (US)

Blumea
Blumea (Netherlands)

Bost. Stud. Philos. Sci.
Boston Studies in the Philosophy of Science (US)

Bot. Abh.
Botanische Abhandlungen aus dem Gebiet der Morphologie und Physiology (Germany)

Bot. Arch.
Botanisches Archiv (Germany)

Bot. Gaz.
Botanical Gazette (US)

Bot. J. Linn. Soc.
Botanical Journal of the Linnean Society (England)

Bot. Rev.
Botanical Review (US)

Bot. Tidsskr.
Botanisk Tidsskrift (Denmark)

Bot. Zblt.
Botanisches Zentralblatt (Germany)

Brain Behav. Evol.
Brain Behavior and Evolution (Switzerland)

Brit. Birds
British Birds (England)

Brit. Ecol. Soc. Symp.
British Ecological Society Symposium (England)

Brit. J. Hist. Sci.
British Journal of the History of Science (England)

Brit. J. Philos. Sci.
British Journal of Philosophy of Science (England)

Brit. Med. Bull.
British Medical Bulletin (England)

Brit. Med. J.
British Medical Journal (England)

Brittonia
Brittonia (US)

Brookh. Symp. Biol.
Brookhaven Symposia in Biology (US)

Bull. Acad. Imp. Sci. St. Petersbourg
Bulletin de l'Académie Impériale des Sciences de Saint Petérsbourg (Russia)

Bull. Acad. Roy. Med., Paris
Bulletin de l'académie royale de medicine, Paris (France)

Bull. Amer. Mus. Nat. Hist.
Bulletin of the American Museum of Natural History (US)

Bull. Anim. Behav.
Bulletin of Animal Behaviour (England)

Bull. Biol. Fr. Belg.
Bulletin Biologique de la France et de la Belgique (France)

Bull. Bot. Soc. Bengal
Bulletin of the Botanical Society of Bengal (India)

Bull. Brit. Mus. Hist. Ser.
Bulletin of the British Museum (Natural History) Historical Series (England)

Bull. Entomol. Res.
Bulletin of Entomological Research (England)

Bull. Geol. Soc. Amer.
Bulletin of the Geological Society of America (US)

Bull. Hist. Med.
Bulletin of the History of Medicine (US)

Bull. Ill. St. Lab. Nat. Hist.
Bulletin of the Illinois State Laboratory of Natural History (US)

Bull. Math. Biophys.
Bulletin of Mathematical Biophysics (US)

Bull. Soc. Belg. Geol. Paleontol. Hydrol.
Bulletin de la Societe Belge de Geologie de Paleontologie et d'Hydrologie (Belgium)

Bull. Soc. For. Franche-Comté
Bulletin de la Societé Forestière de Franche-Comté et Belfort (France)

Bull. Soc. Vaud. Sci. Nat.
Bulletin de la Societe Vaudoise des Sciences Naturelles (Switzerland)

Bull. Torr. Bot. Club
Bulletin of the Torrey Botanical Club (US)

Bull. Univ. Nebr. State Mus.
Bulletin of the University of Nebraska State Museum (US)

C.R. Acad. Sci.
Comptes Rendus Academie des Sciences (France)

C.R. Trav. Lab. Carlsb.
Compte Rendu des Travaux du Laboratoire de Carlsberg (Denmark)

Can. Entomol.
Canadian Entomologist (Canada)

Can. J. Bot.
Canadian Journal of Botany (Canada)

Can. J. Genet. Cytol.
Canadian Journal of Genetics and Cytology (Canada)

Can. J. Zool.
Canadian Journal of Zoology (Canada)

Cancer Res.
Cancer Research (US)

Carnegie Inst. Wash. Publ.
Carnegie Institute of Washington Publication (US)

Cellule
La Cellule (Belgium)

Centaurus
Centaurus (Denmark)

Chem. Rev.
Chemical Reviews (US)

Chromosoma
Chromosoma (Germany)

Chron. Bot.
Chronica Botanica (US)

Clio Med.
Clio Medica (Netherlands)

Cold Spr. Harb. Symp. Quant. Biol.
Cold Spring Harbor Symposia on Quantitative Biology (US)

Condor
Condor (US)

Contemp. Top. Immunobiol.
Contemporary Topics in Immunobiology (US)

Curr. Top. Dev. Biol.
Current Topics in Developmental Biology (US)

Curr. Top. Pathol.
Current Topics in Pathology (Germany)

Cytologia
Cytologia (Japan)

Daedalus
Daedalus (US)

Dana-Rep.
Dana-Report Carlsberg Foundation (Denmark)

Deut. Med. Woch.
Deutsche Medizinische Wochenschrift (Germany)

Dev. Biol.
Developmental Biology (US)

Disc. Faraday Soc.
Discussions of the Faraday Society (England)

Discovery
Discovery (England)

Ecol. Monogr.
Ecological Monographs (US)

Ecology
Ecology (US)

Econ. Bot.
Economic Botany (US)

Endeavour
Endeavour (England)

Entomol. Exp. Appl.
Entomologia Experimentalis et Applicata (Netherlands)

Ergeb. Biol.
Ergebnisse der Biologie (Germany)

Eur. J. Biochem.
European Journal of Biochemistry (Germany)

Eur. J. Immunol.
European Journal of Immunology (Germany)

Evol. Biol.
Evolutionary Biology (US)

Evol. Theory
Evolutionary Theory (US)

Evolution
Evolution (US)

Exp. Cell Res.
Experimental Cell Research (US)

Exp. Gerontol.
Experimental Gerontology (England)

Exp. Parasit.
Experimental Parasitology (US)

Experientia
Experientia (Switzerland)

Fed. Proc.
Federation Proceedings (US)

Folia Primatol.
Folia Primatologica (Switzerland)

Forma Functio
Forma et Function (Germany)

Gen. Syst.
General Systems (US)

Genet. Res.
Genetical Research (England)

Genetics
Genetics (US)

Geol. Soc. Amer. Mem.
Geological Society of America Memoir (US)

Geology
Geology (US)

Growth
Growth (US)

Hereditas
Hereditas (Sweden)

Heredity
Heredity (England)

Hilgardia
Hilgardia (US)

Hist. Sci.
History of Science (England)

Hoppe-Seyl. Z.
Hoppe-Seyler's Zeitschrift für Physiologische Chemie (Germany)

Hum. Biol.
Human Biology (US)

Ibis
Ibis (England)

Icarus
Icarus (US)

Insectes Sociaux
Insectes Sociaux (France)

Int. Rev. Cytol.
International Review of Cytology (US)

Isis
Isis (US)

Isr. J. Bot.
Israel Journal of Botany (Israel)

Jahrb. Wiss. Bot.
Jahrbuch für Wissenschaftliche Botanik (Germany)

Jap. J. Ecol.
Japanese Journal of Ecology (Japan)

Jap. J. Genet.
Japanese Journal of Genetics (Japan)

Jena. Z. Naturw.
Jenaische Zeitschrift für Naturwissenschaft (Germany)

J. Acad. Nat. Sci. Phila.
Journal of the Academy of Natural Sciences of Philadelphia (US)

J. Agric. Res.
Journal of Agricultural Research (US)

J. Amer. Chem. Soc.
Journal of the American Chemical Society (US)

J. Amer. Oil Chem. Soc.
Journal of the American Oil Chemists' Society (US)

J. Amer. Soc. Agron.
Journal of the American Society of Agronomy (US)

J. Anim. Ecol.
Journal of Animal Ecology (England)

J. Anim. Sci.
Journal of Animal Science (US)

J. Arnold Arbor.
Journal of the Arnold Aboretum of Harvard University (US)

J. Bacteriol.
Journal of Bacteriology (US)

J. Biol. Chem.
Journal of Biological Chemistry (US)

J. Biomech.
Journal of Biomechanics (US)

J. Cell. Comp. Physiol.
Journal of Cellular and Comparative Physiology (US)

J. Cell Sci1
Journal of Cell Science (England)

J. Chem. Educ.
Journal of Chemical Education (US)

J. Comp. Physiol. Psychol.
Journal of Comparative and Physiological Psychology (US)

J. Ecol.
Journal of Ecology (England)

J. Econ. Entomol.
Journal of Economic Entomology (US)

J. Embryol. Exp. Morphol.
Journal of Embryology and Experimental Morphology (England)

J. Environ. Manage.
Journal of Environmental Management (England)

J. Exp. Med.
Journal of Experimental Medicine (US)

J. Exp. Zool.
Journal of Experimental Zoology (US)

J. Fish. Res. Bd. Can.
Journal of the Fisheries Research Board of Canada (Canada)

J. Gen. Microbiol.
Journal of General Microbiology (England)

J. Gen. Physiol.
Journal of General Physiology (US)

J. Genet.
Journal of Genetics (India)

J. Geol.
Journal of Geology (US)

J. Geront.
Journal of Gerontology (US)

J. Hered.
Journal of Heredity (US)

J. Hist. Biol.
Journal of the History of Biology (US)

J. Hist. Ideas
Journal of the History of Ideas (US)

J. Hist. Med.
Journal of the History of Medicine and Allied Sciences (US)

J. Immunol.
Journal of Immunology (US)

J. Kans. Entomol. Soc.
Journal of the Kansas Entomological Society (US)

J. Laryngol. Otol.
Journal of Laryngology and Otology (England)

J. Linn. Soc. Lond. Bot.
Journal of the Linnean Society of London Botany (England)

J. Linn. Soc. Lond. Zool.
Journal of the Linnean Society of London Zoology (England)

J. Mammal.
Journal of Mammalogy (US)

J. Med. Genet.
Journal of Medical Genetics (England)

J. Mol. Biol.
Journal of Molecular Biology (US)

J. Mol. Evol.
Journal of Molecular Evolution (Germany)

J. Morph.
Journal of Morphology (US)

J. N.Y. Entomol. Soc.
Journal of the New York Entomological Society (US)

J. Natl. Cancer Inst.
Journal of the National Cancer Institute (US)

J. Ornithol.
Journal für Ornithologie (Germany)

J. Parasit.
Journal of Parasitology (US)

J. Physiol.
Journal of Physiology (England)

J. Protozool.
Journal of Protozoology (US)

J. Reprod. Fertil.
Journal of Reproduction and Fertility (England)

J. Roy. Hort. Soc.
Journal of the Royal Horticultural Society (England)

J. St. Med.
Journal of State Medicine (England)

J. Theor. Biol.
Journal of Theoretical Biology (England)

J. Wash. Acad. Sci.
Journal of the Washington Academy of Sciences (US)

J. Wildl. Manage.
Journal of Wildlife Management (US)

J. Ver. Vaterl. Naturk. Württ.
Jahresheft des Vereins für Vaterländische Naturkunde in Württemberg (Germany)

K. Sven. Vetenskapsakad. Handl.
Kunglica Svenska Vetenskapsakademiens Handlingar (Sweden)

Kew Bull.
Kew Bulletin (England)

Klin. Jahrb.
Klinisches Jahrbuch (Germany)

Klin. Wochenschr.
Klinische Wochenschrift (Germany)

Lethaia
Lethaia (Norway)

Limnol. Oceanogr.
Limnology and Oceanography (US)

Living Bird
Living Bird (US)

Lunds Univ. Arssk.
Lunds Universitets Årsskrift (Sweden)

Madroño
Madroño (US)

Man
Man: The Journal of the Royal Anthropological Institute (England)

Mar. Biol. (Berlin & N.Y.)
Marine Biology (Germany & US)

Math. Biosci.
Mathematical Biosciences (US)

Med. Hist.
Medical History (England)

Mem. Acad. Roy. Sci. Lett. Belg.
Mémoires de l'Académie Royal des Sciences, des Lettres et des Beaux-arts de Belgique (Belgium)

Mem. Acad. Sci.
Mémoires de l'Académie des Sciences de l'Institut de France (France)

Mem. Amer. Acad. Arts. Sci.
Memoirs of the American Academy of Arts and Sciences (US)

Mem. Boston Soc. Nat. Hist.
Memoirs of the Boston Society of Natural History (US)

Mem. Geol. Surv. U.K.
Memoirs of the Geological Survey of the United Kingdom (England)

Mol. Cell. Biochem.
Molecular and Cellular Biochemistry (Netherlands)

Mol. Gen. Genet.
Molecular and General Genetics (Germany)

Mont. Agri. Exp. Sta., Bozeman, Circ.
Montana Agricultural Experiment Station, Bozeman, Circular (US)

Münch. Med. Wochens.
Münchener Medizinische Wochenschrift (Germany)

Mutat. Res.
Mutation Research (Netherlands)

N.Y. State Mus. Bull.
New York State Museum Bulletin (US)

Nat. New Biol.
Nature New Biology (England)

Nat. Sci.
Natural Science (England)

Natl. Cancer Inst. Monogr.
National Cancer Institute Monograph (US)

Naturalist, Hull
Naturalist Hull, London (England)

Nature
Nature (England)

Naturw. Rdsch.
Naturwissenschaftliche Rundschau (Germany)

Naturwissenschaften
Naturwissenschaften (Germany)

New Biol.
New Biology (England)

New Phytol.
New Phytologist (England)

New Sci.
New Scientist (England)

North Amer. Fauna
North American Fauna (US)

Northwest Sci.
Northwest Science (US)

Not. Nat.
Notulae Naturae (US)

Nov. Comm. Acad. Sci. Imp. Petrop.
Novi Comentarii Academiae Scientiarum Imperalis Petropolitanae (Russia)

Nova Acta Leopold.
Nova Acta Leopoldina (Germany)

Occas. Pap. Mus. Zool. Univ. Mich.
Occasional Papers of the Museum of Zoology University of Michigan (US)

Oecologia
Oecologia (Germany)

Orig. Life
Origins of Life (Netherlands)

Österr. Bot. Z.
Österreichische Botanische Zeitschrift (Austria)

Pac. Insects Monogr.
Pacific Insects Monograph (US)

Paleobiology
Paleobiology (US)

Parasitology
Parasitology (England)

Perspect. Biol. Med.
Perspectives in Biology and Medicine (US)

Pflanz.-Nach. Bayer
Pflanzenschutz-Nachrichten Bayer (Germany)

Pflügers Arch. Ges. Physiol.
Pflügers Archiv für die Gesamte Physiologie der Menschen und der Tiere (Germany)

Phil. Trans. Roy. Soc. Lond.
Philosophical Transactions of the Royal Society of London (England)

Phil. Trans. Roy. Soc. Lond. A
Philosophical Transactions of the Royal Society of London, Series A Mathematical and Physical Sciences (England)

Phil. Trans. Roy. Soc. Lond. B
Philosophical Transactions of the Royal Society of London, Series B Biological Sciences (England)

Philos. Sci.
Philosophy of Science (US)

Philosophy
Philosophy (England)

Photochem. Photobiol.
Photochemistry and Photobiology (England)

Physiol. Rev.
Physiological Reviews (US)

Physiol. Zool.
Physiological Zoology (US)

Phytomorphology
Phytomorphology (India)

Phyton
Phyton (Austria)

Phytopathol. Z.
Phytopathologische Zeitschrift (Germany)

Phytopathology
Phytopathology (US)

Plant Physiol.
Plant Physiology (US)

Planta
Planta (Germany)

Poggendorffs Ann.
Poggendorffs Annalen (Germany)

Pop. Sci. Mon.
Popular Science Monthly (US)

Proc. Acad. Nat. Sci. Phila.
Proceedings of the Academy of Natural Sciences of Philadelphia (US)

Proc. Acad. Sci. Armenian SSR
Proceedings of the Academy of Sciences of the Armenian S.S.R.

Proc. Amer. Acad. Arts Sci.
Proceedings of the American Academy of Arts and Sciences (US)

Proc. Amer. Philos. Soc.
Proceedings of the American Philosophical Society (US)

Proc. Berkeley Symp. Math. Stat. Probab.
Proceedings of the Berkeley Symposium on Mathematical Statistics and Probability (US)

Proc. Cal. Acad. Sci.
Proceedings of the California Academy of Sciences (US)

Proc. Camb. Philos. Soc.
Proceedings of the Cambridge Philosophical Society (England)

Proc. Ecol. Soc. Aust.
Proceedings of the Ecological Society of Australia (Australia)

Proc. Indiana Acad. Sci.
Proceedings of the Indiana Academy of Science (US)

Proc. Int. Bot. Congr.
Proceedings of the International Botanical Congress (International)

Proc. Int. Congr. Genet.
Proceedings of the International Congress of Genetics (Italy)

Proc. Int. Congr. Hist. Sci.
Proceedings of the International Congress of the History of Science (International)

Proc. Int. Ornithol. Congr.
Proceedings International Ornithological Congress (International)

Proc. Jap. Acad.
Proceedings of the Japan Academy (Japan)

Proc. Natl. Acad. Sci.
Proceedings of the National Academy of Sciences of the United States of America (US)

Proc. Roy. Entomol. Soc. Lond.
Proceedings of the Royal Entomological Society of London (England)

Proc. Roy. Entomol. Soc. Lond. C
Proceedings of the Royal Entomological Society of London Series C Journal of Meetings (England)

Proc. Roy. Soc. Edinb.
Proceedings of the Royal Society of Edinburgh (Scotland)

Proc. Roy. Soc. Edinb. B
Proceedings of the Royal Society of Edinburgh Series B Biology (Scotland)

Proc. Roy. Soc. Lond.
Proceedings of the Royal Society of London (England)

Proc. Roy. Soc. Lond. A
Proceedings of the Royal Society of London, Series A: Mathematical and Physical Sciences (England)

Proc. Roy. Soc. Lond. B
Proceedings of the Royal Society of London, Series B: Biological Sciences (England)

Proc. Roy. Soc. Vict.
Proceedings of the Royal Society of Victoria (Australia)

Proc. Zool. Soc. Lond.
Proceedings of the Zoological Society of London (England)

Prof. Geogr.
Professional Geographer (US)

Prog. Allergy
Progress in Allergy (Switzerland)

Prog. Biophys. Biophys. Chem.
Progress in Biophysics and Biophysical Chemistry (England)

Prog. Biophys. Mol. Biol.
Progress in Biophysics and Molecular Biology (England)

Prog. Exp. Tumor Res.
Progress in Experimental Tumor Research (Switzerland)

Prog. Nucl. Acid. Res.
Progress in Nucleic Acid Research (US)

Prog. Theor. Biol.
Progress in Theoretical Biology (US)

Protoplasma
Protoplasma (Austria)

Psyche
Psyche (US)

Psychol. Rev.
Psychological Review (US)

Q. J. Microsc. Sci.
Quarterly Journal of Microscopical Science (England)

Q. Rev. Biol.
Quarterly Review of Biology (US)

Q. Rev. Biophys.
Quarterly Review of Biophysics (England)

Rep. Brit. Assoc. Adv. Sci.
Report of the British Association for the Advancement of Science (England)

Rep. Evol. Comm. Roy. Soc. Lond.
Report to the Evolution Committee of the Royal Society of London (England)

Rev. Ecol. Biol. Sol
Revue d'Ecologie et Biologie du Sol (France)

Rev. Gén. Bot.
Revue Génerale de Botanique (France)

Rev. Hist. Sci. Applic.
Revue d'Histoire des Sciences et de Leurs Applications (France)

Rev. Mod. Phys.
Review of Modern Physics (US)

Sat. Rev. Lit.
Saturday Review of Literature (US)

Sber. K. Böhm. Ges. Wiss.
Sitzungsberichte der Koniglich Böhmischen Gesellschaft der Wissenschaften (Czechoslovakia)

Sber. K. Preuss. Akad. Wiss.
Sitzungsberichte der Königlich Preussischen Akademie der Wissenschaften (Germany)

Sci. Amer.
Scientific American (US)

Sci. Mon.
Scientific Monthly (US)

Sci. Prog.
Science Progress (England)

Science
Science (US)

Scientia
Scientia (Italy)

Search
Search (Australia)

Smithson. Inst. Annu. Rep.
Smithsonian Institution Annual Report (US)

Smithson. Misc. Coll.
Smithsonian Miscellaneous Collections (US)

Stadler Genet. Symp.
Stadler Genetics Symposia (US)

Stud. Hist. Philos. Sci.
Studies in History and Philosophy of Science (England)

Surv. Biol. Prog.
Survey of Biological Progress (US)

Svensk Bot. Tidskr.
Svensk Botanisk Tidskrift (Sweden)

Symp. Roy. Entomol. Soc. Lond.
Symposia of the Royal Entomological Society of London (England)

Symp. Soc. Exp. Biol.
Symposia of the Society for Experimental Biology (US)

Symp. Soc. Gen. Microbiol.
Symposium of the Society for General Microbiology (England)

Symp. Zool. Soc. Lond.
Symposia of the Zoological Society of London (England)

Syst. Assoc. Publ.
Systematics Association Publication (England)

Syst. Bot.
Systematic Botany (US)

Syst. Zool.
Systematic Zoology (US)

Taxon
Taxon (Netherlands)

Texas Agric. Exp. Sta. Bull.
Texas Agricultural Experiment Station Bulletin (US)

Theor. Appl. Genet.
Theoretical and Applied Genetics (Germany)

Theor. Pop. Biol.
Theoretical Population Biology (US)

Topology
Topology (England)

Torreya
Torreya (US)

Trans. Amer. Philos. Soc.
Transactions of the American Philosophical Society (US)

Trans. Conn. Acad. Arts Sci.
Transactions Connecticut Academy of Arts and Sciences (US)

Trans. Dynam. Dev.
Transactions on the Dynamics of Development (Russia)

Trans. Linn. Soc. Lond.
Transactions of the Linnean Society of London (England)

Trans. Roy. Entomol. Soc. Lond.
Transactions of the Royal Entomological Society of London (England)

Trans. Roy. Soc. Can.
Transactions of the Royal Society of Canada (Canada)

Trans. Roy. Soc. Edinb.
Transactions of the Royal Society of Edinburgh (Scotland)

Transplant. Rev.
Transplantation Reviews (Denmark)

Trudy Azerbajdzansk. Opytn. Sta.
Trudy Azerbajdzanskogo Opytnoj Stancii (Russia)

U.S.D.A. Bur. Biol. Surv. Bull.
U.S. Department of Agriculture. Bureau of Biological Survey. Bulletin (US)

U.S.D.A. Misc. Publ.
U.S. Department of Agriculture. Miscellaneous Publications (US)

U.S.D.A. Tech. Bull.
U.S. Department of Agriculture. Technical Bulletin (US)

U.S.D.A. Yearb. Agric.
U.S. Department of Agriculture. The Yearbook of Agriculture (US)

Univ. Chicago Publ.
University of Chicago Publications (US)

Vegetatio
Vegetatio (Netherlands)

Verh. Natur.-Med. Ver. Heidelb.
Verhandlungen des Naturhistorisch-Medizinischen Vereins du Heidelberg (Germany)

Verh. Naturforsch. Ver. Brünn
Verhandlungen des Naturforschenden Vereins in Brünn (Germany)

Verh. Phys.-Med. Ges. Würzburg
Verhandlungen der Physikalisch-Medicinischen Gesellschaft in Würzburg (Germany)

Versl. Landbouwk. Onderz.
Verslagen van Landbouwkundige Onderzoekingen (Netherlands)

Viert. Naturf. Ges. Zürich
Vierteljahrschrift der Naturforschenden Gesellschaft in Zürich (Switzerland)

Virchows Arch. Path. Anat.
Virchows Archiv für Pathologische Anatomie (Germany)

Vopr. Filos.
Voprosy Filosofii (Russia)

Wilhelm Roux' Arch.
Wilhelm Roux' Archiv für Entwicklungsmechanik der Organismen (Germany)

Wilson Bull.
Wilson Bulletin (US)

Z. *Allg. Physiol.*
Zeitschrift für Allgemeine Physiologie (Germany)

Z. *Angew. Ent.*
Zeitschrift für Angewandte Entomologie (Germany)

Z. *Biol.*
Zeitschrift für Biologie (Germany)

Z. *Indukt. Abstammungs-Vererbungsl.*
Zeitschrift für Induktive Abstammungs- und Vererbungslehre (Germany)

Z. *Morphol. Oekol. Tiere*
Zeitschrift für Morphologie und Oekologie der Tiere (Germany)

Z. *Naturforsch. B*
Zeitschrift für Naturforschung, Teil B (Germany)

Z. Parasitenk.
Zeitschrift für Parasitenkunde (Germany)

Z. Pflanzenkr.
Zeitschrift für Pflanzenkrankheiten (Germany)

Z. Pflanzenkr. Pflanzenschutz
Zeitschrift für Pflanzenkrankheiten und Pflanzenschutz (Germany)

Z. Physik
Zeitschrift für Physik (Germany)

Z. Säugetierk.
Zeitschrift für Säugetierkunde (Germany)

Z. Tierpsychol.
Zeitschrift für Tierpsychologie (Germany)

Z. Vergl. Physiol.
Zeitschrift für Vergleichende Physiologie (Germany)

Z. Wiss. Zool.
Zeitschrift für Wissenschaftliche Zoologie (Germany)

Z. Zool. Syst. Evolutionsforsch.
Zeitschrift für Zoologische Systematik und Evolutionsforschung (Germany)

Zbl. Bakter. C. Orig.
Zentralblatt für Bakteriologie Parasitenkunde Infektions-Krankheiten und Hygiene, Abteilung I Originale (Germany)

Zh. Obshch. Biol.
Zhurnal Obshchei Biologii (Russia)

Zool. Ann.
Zoologische Annalen (Germany)

Zool. Anz.
Zoologischer Anzeiger (Germany)

Zool. Beitr.
Zoologische Beiträge (Germany)

Zool. Bidr. Upps.
Zoologiska Bidrag Fran Uppsala (Sweden)

Zool. Jhrb. Abt. Allg. Zool. Physiol. Tiere
Zoologische Jahrbücher Abteilung für Allgemeine Zoologie und Physiologie der Tiere (Germany)

Zool. Jahrb. Abt. Syst.
Zoologische Jahrbücher Abteilung für Systematik Oekologie und Geographie der Tiere (Germany)

Zool. Pol.
Zoologica Poloniae (Poland)

Zool. Scr.
Zoologica Scripta (Sweden)

Zoologica (N.Y.)
Zoologica (US)

Dictionary of Theoretical Concepts in Biology

[A]

A-RATES *see* **TACHYTELY**

a SELECTION [evolution]

Gill, D.E. 1972. Intrinsic rates of increase, saturation densities, and competitive ability. I. An experiment with *Paramecium*. *Amer. Nat.* 106:461–71.

______ . 1974. Intrinsic rate of increase, saturation density, and competitive ability. II. The evolution of competitive ability. *Amer. Nat.* 108:103–16.

Pianka, E.R. 1972. r and K selection or b and d selection? *Amer. Nat.* 106:581–88.

Vandermeer, J.H. 1972. On the covariance of the community matrix. *Ecology* 53:187–89.

ABEL'S LAW *see* **BIOLOGICAL MOMENTUM, LAW OF**

ABIOGENESIS

Oparin, A.I. 1953. *The Origin of Life.* 2d ed. Transl. by S. Morgulis. New York: Dover.

Haldane, J.B.S. 1954. The origins of life. *New Biol.* 16:12–27.

Calvin, M. 1956. Chemical evolution and the origin of life. *Amer. Sci.* 44:248–63.

Cloud, P.E., Jr. 1968. Pre-Metozoan evolution and the origins of the Metozoa. *Evolution and Environment.* Peabody Museum Centennial Symposium, Yale University, 1966. New Haven, Conn.: Yale University Press, pp. 5–7.

Cf. SPONTANEOUS GENERATION; HETEROGENESIS

ABIOSIS *see* **ANABIOSIS**

 ABOMINABLE MYSTERY *see* **DARWIN'S ABOMINABLE MYSTERY**

ABRUPT-ECOSPECIES [evolution]
Valentine, D.H. 1949. The units of experimental taxonomy. *Acta Biotheor*. 9:75–88.
Jones, K. 1961. The status and development of abrupt-ecospecies. *Recent Advances in Botany*—2 vols., Toronto: University of Toronto Press. I:862–66.

ACCEPTER SPECIES [bird behavior]
Rothstein, S.I. 1975. Evolutionary rates and host defenses against avian brood parasitism. *Amer. Nat.* 109:161–76.
______. 1975. An experimental and teleonomic investigation of avian brood parasitism. *Condor* 77:250–71.
Cf. REJECTOR SPECIES

ACCESSIBILITY [ecology]
Kellmann, M.C. 1970. The influence of accessibility on the composition of vegetation. *Prof. Geogr*. 22:1–4.

ACCUMULATION OF RANDOM MUTATIONS, THEORY OF [genetics]
= PANMIXIA
Barr, T.C., Jr. 1968. Cave ecology and the evolution of Troglobites. *Evol. Biol*. 2:35–102.
Wilkens, H. 1971. Genetic interpretation of regressive evolutionary processes: studies on hybrid eyes of two astyanax cave populations *(Characidae, Pisces)*. *Evolution* 25:530–44.

ACQUIRED CHARACTERISTICS [evolution]
Lankester, E.R. 1894. Acquired characters. *Nature* 51:102.
Zirkle, C. 1946. The early history of the idea of the inheritance of acquired characters and of pangenesis. *Trans. Amer. Philos. Soc*. 35:91–151.
Koestler, A. 1971. *The Case of the Midwife Toad*. New York: Random House.
Cf. LAMARCKIAN EVOLUTION; PANGENESIS

ACTIVE CENTER *see* **REACTIVE SITE**

ACTIVE TRANSPORT [cell physiology]
Fick, A.E. 1855. Uber diffusion. *Poggendorffs Ann*. 94:59–86.
Society for Experimental Biology. 1954. *Active Transport and Secretion: Symposium number VIII*. New York: By the Society.

Murphy, Q.R., ed. 1957. *Metabolic Aspects of Transport Across Cell Membranes*. Madison: University of Wisconsin Press (esp. H. H. Ussing: General principles and theories of membrane transport, pp. 39-56).

Stein, W.D. 1967. *The Movement of Molecules Across Cell Membranes*. New York: Academic Press.

Sarkadi, B., & D.C. Tosteson. 1978. Active cation transport in human red cells. *Membrane Transport in Biology*. vol 2: *Transport Across Single Biological Membranes*. Edited by D.C. Tosteson. Berlin: Springer-Verlag, pp. 117–60.

ADAPTATION [evolution]

Stern, J.T., Jr. 1970. The meaning of "adaptation" and its relation to the phenomenon of natural selection. *Evol. Biol.* 4:39–66.

Munson, R. 1976. Biological adaptation. *Bost. Stud. Philos. Sci.* 27:330–50.

Cf. FITNESS; TELEONOMY

ADAPTATION TOLERANCE [host-parasite interaction]

Sprent, J. F. A. 1959. Parasitism, immunity and evolution. *The Evolution of Living Organisms*. A Symposium of the Royal Society, Victoria, pp. 149–65.

Noble, E. R., & G. A. Noble. 1976. *Parasitology: The Biology of Animal Parasites*. 4th ed. Philadelphia: Lea & Febiger, pp. 507–08.

ADAPTEDNESS [evolution]

Dobzhansky, T. 1956. What is an adaptive trait? *Amer. Nat.* 90:337–47.

______ . 1968. Adaptedness and fitness. *Population Biology and Evolution*. Edited by R. C. Lewontin. Syracuse, N.Y.: Syracuse University Press, pp. 109–21.

ADAPTEDNESS-ADAPTABILITY BALANCE HYPOTHESIS [evolution]

Mather, K. 1953. Genetical structure of populations. *Symp. Soc. Exp. Biol.* 7:66–95.

______ . 1973. *Genetical Structure of Populations*. London: Chapman and Hall, chap. 7.

______ . , & D. deWinton. 1941. Adaptation and counter-adaptation of the breeding system in *Primula*. *Ann. Bot.* 5:297–311.

Jain, S. K. 1976. The evolution of inbreeding in plants. *Annu. Rev. Ecol. Syst.* 7:469–95.

ADAPTIVE INDEX [evolution]

Frazzetta, T.H. 1975. *Complex Adaptations in Evolving Populations*. Sunderland. Mass.: Sinauer Associates, chap 3.

ADAPTIVE LANDSCAPE [evolution]

= ADAPTIVE ZONES

McCoy, J. W. 1979. The origin of the "adaptive landscape" concept. *Amer. Nat.* 113:610–13.

Cf. ADAPTIVE PEAK

ADAPTIVE MODIFICATION ALONG THE LINES OF LEAST RESISTANCE [evolution]

Ganong, W. F. 1901. The cardinal principles of morphology. *Bot. Gaz.* 31:426–34.

Stebbins, G. L. 1950. *Variation and Evolution in Plants*. New York: Columbia University Press, p. 497.

______. 1974. *Flowering Plants: Evolution Above the Species Level*. Cambridge: Belknap Press of Harvard University Press, p. 31.

ADAPTIVE NEUTRALITY *see* **NEO-DARWINIAN EVOLUTION; NEUTRAL THEORY**

ADAPTIVE PEAK [population fitness]

= Wright's principle of maximization of W = adaptive foci

Wright, S. 1931. Evolution in Mendelian populations. *Genetics* 16:97–159.

______. 1932. The roles of mutation, inbreeding, crossbreeding and selection in evolution. *Proc. 6th Int. Congr. Genet.* 1:356–66.

______. 1955. Classification of the factors of evolution. *Cold Spr. Harb. Symp. Quant. Biol.* 20:16–24.

______. 1969. *Evolution and the Genetics of Populations*. vol. 2: *Theory of Gene Frequencies*. Chicago: University of Chicago Press, chap. 5.

Cf. FITNESS [Darwinian]

ADAPTIVE RADIATION

Osborn, H. F. 1902. The law of adaptive radiation. *Amer. Nat.* 36:353–63.

Simpson, G. G. 1953. *The Major Features of Evolution*. New York: Columbia University Press, chap. 7.

Stebbins, G. L. 1967. Adaptive radiation and trends of evolution in higher plants. *Evol. Biol.* 1:101–42.

______. 1971. Relationships between adaptive radiation, speciation and major evolutionary trends. *Taxon* 20:3–16.

ADAPTIVE SURFACE [evolution]

Wright, S. 1967. "Surfaces" of selective value. *Proc. Natl. Acad. Sci.* 58:165–72.

Turner, J. R. G. 1971. Wright's adaptive surface, and some general rules for equilibria in complex polymorphisms. *Amer. Nat.* 105:267–78.

ADAPTIVE TOPOGRAPHY [evolution]

Wright, S. 1932. The roles of mutation, inbreeding, crossbreeding and selection in evolution. *Proc. 6th Int. Congr. Genet.*, 1:356–66.

———. 1967. "Surfaces" of selective value. *Proc. Natl. Acad. Sci.* 58:165–72.

Moran, P. A. P. 1964. The nonexistence of adaptive topographies. *Ann. Hum. Genet.* 27:383–93.

Cf. ADAPTIVE LANDSCAPE; ADAPTIVE PEAK; ADAPTIVE SURFACE; ADAPTIVE ZONES; TWO LOCUS THEORY

ADAPTIVE ZONES [evolution]

Wright, S. 1931. Evolution in Mendelian populations. *Genetics* 16:97–159.

Simpson, G. G. 1953. *The Major Features of Evolution.* New York: Columbia University Press, p. 199*ff.*

Van Valen, L. 1971. Adaptive zones and the orders of mammals. *Evolution* 25:420–28.

Cf. ADAPTIVE PEAK

ADAPTOR HYPOTHESIS [protein synthesis]

Crick, F. H. C. 1958. On protein synthesis. *Symp. Soc. Exp. Biol.* 12:138–63.

———. 1963. The recent excitement in the coding problem. *Prog. Nucl. Acid Res.* 1:163–217.

Hoagland, M. B. 1960. The relationship of nucleic acid and protein synthesis as revealed by studies in cell-free systems. *The Nucleic Acids.* Edited by E. Chargaff & J. N. Davidson. 3 vols.: New York: Academic Press, III, 349–408.

ADENOSINE TRIPHOSPHATE (ATP) HYPOTHESIS [of active transport]

Whittam, R. 1964. The interdependence of metabolism and active transport. *The Cellular Functions of Membrane Transport.* Edited by J. E. Hoffman. Englewood Cliffs, N.J.: Prentice-Hall, pp. 139–54.

ADJUSTMENT STABILITY [ecology]

Margalef, R. 1969. Diversity and stability: a practical proposal and a model of interdependence. *Brookh. Symp. Biol.* 22:25–37.

Conwell, J. H., & R. O. Slayter. 1977. Mechanisms of succession in natural communities and their role in community stability and organization. *Amer. Nat.* 111:1119–44.

Cf. SUCCESSION

ADVERGENCE [evolution]

Brower, L. P., & J. van Z. Brower. 1972. Parallelism, convergence, divergence, and the new concept of advergence in the evolution of mimicry. *Trans. Conn. Acad. Arts Sci.* 44:57–67.

AERIAL PLANKTON IDEA [insect distribution]

= aeroplankton

Coad, B. R. 1931. Insects captured by air plane are found at surprising heights. *U.S.D.A. Yearb. Agric.* 1931:320–23.

Hardy, A. C., & P. S. Milne. 1937. Insect drift over the North Sea. *Nature* 139:510–11.

Glick, P. A. 1939. The distribution of insects, spiders, and mites in the air. *U.S.D.A. Tech. Bull.* 673.

Johnson, C.G. 1969. *Migration and Dispersal of Insects by Flight.* London: Methuen, chap. 13.

AEROPLANKTON *see* **AERIAL PLANKTON IDEA**

AGAMIC COMPLEX [evolution]

Babcock, E. B., & G. L. Stebbins. 1938. The American species of *Crepis:* their interrelationships and distribution as affected by polyploidy and apomixis. *Carnegie Inst. Wash. Publ.* 504.

Dobzhansky, T., F. J. Ayala, G. L. Stebbins, & J. W. Valentine. *Evolution.* San Francisco: Freeman, chap. 7.

Cf. BLENDING INHERITANCE

AGE AND AREA [biogeography]

Berry, E. W. 1917. A note on the "age and area" hypothesis. *Science* 46:539–40.

Arber, A. 1919. On the law of age and area, in relation to the extinction of species. *Ann. Bot.* 33:211–14.

Willis, J. C. 1922. *Age and Area.* Cambridge: Cambridge University Press.

Wright, S. 1941. The "age and area" concept extended. *Ecology* 22:345–47.

Stebbins, G. L., Jr. 1942. The genetic approach to problems of rare and endemic species. *Madroño* 6:241–58.

Good, R. 1974. *The Geography of the Flowering Plants.* 4th ed. London: Longman, chap. 3.

AGE POLYETHISM *see* **POLYETHISM**

AGE-SPECIFIC SELECTION [in populations]

Anderson, W. W., & E. C. King. 1970. Age-specific selection. *Proc. Natl. Acad. Sci.* 66:780–86.

King E. C., & W. W. Anderson. 1971. Age-specific selection. II. The interaction between r and K during population growth. *Amer. Nat.* 105:137–56.

AGGRESSION [behavior]

Kaufman, H. 1970. *Aggression and Altruism.* New York: Holt, Rinehart, & Winston.

Johnson, R. N. 1973. *Aggression in Man and Animals.* Philadelphia: Saunders.

Wilson, E. O. 1975. *Sociobiology.* Cambridge: Belknap Press of Harvard University Press, chap. 11.

Cf. AGONISTIC BEHAVIOR

AGGRESSION-DISPERSAL HYPOTHESIS [behavior]

Christian, J. J. 1970. Social subordination, population density, and mammalian evolution. *Science* 168:84–90.

Bekoff, M. 1977. Mammalian dispersal and the ontogeny of individual behavioral phenotypes. *Amer. Nat.* 111:715–32.

AGING

Medawar, P. B. 1957. *The Uniqueness of the Individual.* London: Methuen, chap. 2.

Burnet, F. M. 1973. *The Biology of Ageing.* Auckland: Auckland University Press/Oxford University Press.

Leopold, A. C. 1975. Aging, senescence and turnover in plants. *Bioscience* 25:659–62.

Cf. SENESCENCE

AGONISTIC BEHAVIOR

Scott, J. P., & E. Fredericson. 1951. The causes of fighting in mice and rats. *Physiol. Zool.* 24:273–309.

Hinde, R. A. 1970. *Animal Behavior: A Synthesis of Ethology and Comparative Psychology.* 2d ed. New York: McGraw-Hill, pp. 333–49.

Wilson, E. O. 1975. *Sociobiology.* Cambridge: Belknap Press of Harvard University Press, pp. 242*ff.*

Cf. AGGRESSION

AGONISTIC BUFFER [behavior]

Deag, J. M., & J. H. Crook. 1971. Social behavior and "agonistic buffering" in the wild Barbary macaque, *Macaca sylvania* L. *Folia Primatol.* 15:183–200.

AIR MASS HYPOTHESIS [*Drosophila* genetics]

Crumpacker, D. W., J. Pyati, & L. Ehrman. 1977. Ecological genetics and chromosomal polymorphism in Colorado populations of *Drosophila pseudoobscura*. *Evol. Biol.* 10:437–49.

ALARM-DEFENSE SYSTEM [insect chemistry]

Wilson, E. O. 1971. *The Insect Societies*. Cambridge: Harvard University Press, pp. 235*ff*.

______, & F.E. Regnier, Jr. 1971. The evolution of the alarm-defense system in the formicine ants. *Amer. Nat.* 105:279–89.

ALL-OR-NONE LAW [in nerve fibers]

Chandler, W. K., R. Fitzhugh, & K. S. Cole. 1962. Theoretical stability properties of a space-clamped axon. *Biophys. J.* 2:105–27.

Cole, K. S. 1971. Some aspects of electrical studies of the squid giant axon membrane. *Biophysics and Physiology of Excitable Membranes*. Edited by W. J. Edelman, Jr. New York: Van Nostrand Reinhold, pp. 125–42.

ALLAESTHETIC SELECTION [evolution]

Cott, H. B. 1954. Allaesthetic selection and its evolutionary aspects. *Evolution as a Process*. Edited by J. Huxley, A. C. Hardy, & E. B. Ford. London: Allen & Unwin, pp. 47-70.

ALLEE EFFECT [ecology; evolution]

= Allee's principle

Allee, W. C., A. E. Emerson, O. Park, T. Park, & K. P. Schmidt. 1949. *Principles of Animal Ecology*. Philadelphia: Saunders, p. 389*ff*.

Baker, R.R. 1978. *The Evolutionary Ecology of Animal Migration*. New York: Holmes & Meier, p. 70.

Asmussen, M.A. 1979. Density-dependent selection II. The Allee effect. *Amer. Nat.* 114:796-809.

Cf. DENSITY-DEPENDENT SELECTION

ALLELIC SUBSTITUTION HYPOTHESIS [evolution of inbreeding]

Haldane, J. B. S. 1932. *The Causes of Evolution*. London: Harper, chap. 4.

Jain, S. K. 1976. The evolution of inbreeding in plants. *Annu. Rev. Ecol. Syst.* 7:469–95.

ALLELOCHEMICS [ecology]

Whittaker, R. H. 1975. *Communities and Ecosystems.* 2d ed. New York: Macmillan, pp. 253*ff.*

______ . 1970. The biochemical ecology of higher plants. *Chemical Ecology.* Edited by E. Sondheimer & J. B. Simeone. New York: Academic Press, pp. 43–70.

______ ., & P. P. Feeny. 1971. Allelochemics: chemical interactions between species. *Science* 171:750–70.

ALLELOPATHY [plant chemistry]

Candolle, A. P. de. 1832. *Physiologie Vegetale, ou Exposition des Forces et des Fonctions des Vegetaux* . . . . Paris: Bechet Jeune.

Molisch, H. 1937. *Der Einfluss einer Pflanze auf die andere Allelopathie.* Jena: Fischer.

Muller, C. H. 1969. Allelopathy as a factor in ecological process. *Vegetatio* 18:348–57.

Tuky, H. B. 1969. Implications of allelopathy in agricultural plant science. *Bot Rev.* 35:1–16.

Pickett, S. T., & J. M. Baskin. 1973. Allelopathy and its role in the ecology of higher plants. *Biologist* 55:49–73.

Rice, E. L. 1974. *Allelopathy.* New York: Academic Press.

______ . 1979. Allelopathy—an update. *Bot. Rev.* 45:15–109.

ALLEN'S RULE [biogeography]

Allen, J. A. 1877. The influence of physical conditions in the genesis of species. *Radical Review* 1:108–40.

Mayr, E. 1956. Geographical character gradients and climatic adaptation. *Evolution* 10:105–08.

______ . 1965. *Animal Species and Evolution.* Cambridge: Belknap Press of Harvard University Press, chap. 11.

______ . 1970. *Populations, Species, and Evolution.* Cambridge: Belknap Press of Harvard University Press, p. 199.

Rensch, B. 1959. *Evolution Above the Species Level.* New York: Columbia University Press, pp. 39*ff.*

Ray, C. 1960. The application of Bergmann's and Allen's rules to the poikilotherms. *J. Morph.* 106:85–108.

Cf. BERMANN'S RULE; RENSCH'S RULE

ALLOCATION PRINCIPLE [population genetics]

Levins, R. 1968. *Evolution in Changing Environments.* Princeton, N.J.: Princeton University Press, chap. 2.

Coldy, M. L. 1966. A general theory of clutch size. *Evolution* 20:174–84.

______. 1974. Optimization in ecology. *Science* 183:1156–64.

ALLOGENIC INHIBITION [immunology]

Hellström, K. E. 1964. Possible use of H-2 antigens as positive selective markers in heterogeneous cell populations. *Nature* 201:893–95.

______. 1964. Growth inhibition of sarcoma and carcinoma cells of homozygous origin. *Science* 143:477–78.

Möller, E. 1965. Contact-induced cytotoxicity by lymphoid cells containing foreign isoantigens. *Science* 147:873–74, 879.

ALLOMETRY [relative growth]

= HETEROGONY = developmental canalization = differential relative growth

Huxley, J. S. 1924. Constant differential growth-ratios and their significance. *Nature* 114:895–96

______. 1932. *Problems of Relative Growth*. London: Methuen.

______., & G. Teissier. 1936. Terminology of relative growth. *Nature* 137:780–81.

Thompson, D'A. W. 1942. *On Growth and Form*. 2d ed. Cambridge: Cambridge University Press, chaps. 3, 17.

Reeve, E. C. R., & J. S. Huxley. 1945. Some problems in the study of allometric growth. *Essays on Growth and Form, Presented to D'Arcy Wentworth Thompson*. Edited by W. E. LeGros Clark & P. B. Medawar. Oxford: Clarendon Press, pp. 121–56.

Zuckerman, S. 1950. A discussion on the measurement of growth and form. *Proc. Roy. Soc. Lond. B* 137:433–523.

Simpson, G. G. 1953. *The Major Features of Evolution*. New York: Columbia University Press, p. 25.

Nowinski, W. W. 1960. *Fundamental Aspects of Normal and Malignant Growth*. Amsterdam: Elsevier, p. 224.

Gould, S. J. 1966. Allometry and size in ontogeny and phylogeny. *Biol. Rev.* 41:587–640.

______. 1977. *Ontogeny and Phylogeny*. Cambridge: Belknap Press of Harvard University Press, pp. 341*ff*.

Rosen, R. 1967. *Optimality Principles in Biology*. London: Butterworths, p. 71.

ALLOPATRIC SPECIATION

= GEOGRAPHIC SPECIATION

White, M. J. D. 1968. Models of speciation. *Science* 159:1065–70.

Eldridge, N. 1971. The allopatric model and phylogeny in paleozoic invertebrates. *Evolution* 25:156–67.

——— . , & S. J. Gould. 1972. Punctuated equilibria: an alternative to phyletic gradualism. *Models in Paleobiology*. T. J. M. Schopf. San Francisco: Freeman, Cooper, pp. 82–115.

Dodson, M. M., & A. Hallam. 1977. Allopatric speciation and the fold catastrophe. *Amer. Nat.* 111:415–33.

ALLOSTERY, THEORY OF [protein synthesis]

Monod, J., J. Wyman, & J. P. Changeux. 1965. On the nature of allosteric transitions: a plausible model. *J. Mol. Biol.* 12:88–118.

Whitehead, E. 1970. The regulation of enzyme activity and allosteric transitions. *Prog. Biophys. Mol. Biol.* 21:321–97.

ALPHA-SELECTION [genetics]

Clarke, B. 1973. The effect of mutation on population size. *Nature* 242:196–97.

——— . 1973. Mutation and population size. *Heredity* 31: 367–79.

Wallace, B. 1975. Hard and soft selection revisited. *Evolution* 29:465–73.

Cf. HARD AND SOFT SELECTION

ALTERNATION OF GENERATIONS [evolution]

= metagenesis

Steenstrup, J. J. S. 1845. *On the Alternation of Generations*. Transl. by G. Busk. London: Ray Society.

Haeckel, E. 1866. *Generelle Morphologie der Organismen. Allgemeine Grundzüge der Organischen Formwirtschaft, mechanisch begründet durch die von Ch. Darwin reformierte Deszendenztheorie*. Berlin: Reimer.

Hofmeister, W. 1862. *On the Germination, Development and Fructification of the Higher Cryptogamia and on the Fructification of the Coniferae*. London: Robert Hardwicke, for the Ray Society. (Transl. of German original, 1851).

Allsopp, A. 1966. Developmental stages and life histories in the lower green plants. *Trends in Plant Morphogenesis*. Edited by E. G. Cutter. New York: Wiley, pp. 64–87.

Steere, W. C. 1969. A new look at evolution and phylogeny in bryophytes. *Current Topics in Plant Science*. Edited by J. E. Gunckel. New York: Academic Press, pp. 134–43.

Whittier, D. P. 1971. The value of ferns in an understanding of the Alternation of Generations. *Bioscience* 21:225–27.

Roe, K. E. 1975. Origin of the alternation of generations in plants: reconsideration of the traditional theories. *Biolgist* 57:1–13.

Cf. ANTITHETIC THEORY; HOMOLOGOUS THEORY

ALTERNATION OF SPECIES [ecology]

Schaeffer, R., & R. Moreau. 1958. L'alternance des essences. *Bull. Soc. For. Franche-Compté* 29:1–12, 76–84, 277–98.

Fox, J. F. 1977. Alternation and coexistence of tree species. *Amer. Nat.* 111:69–89.

ALTRUISM [behavior]

Haldane, J. B. S. 1932. *The Causes of Evolution*. London: Longmans, p. 138.

______ . 1955. Population genetics. *New Biol.* 18:34–51.

Hamilton, W. D. 1963. The evolution of altruistic behavior. *Amer. Nat.* 97:354–56.

______ . 1964. The genetical evolution of social behaviour. I and II. *J. Theor. Biol.* 7:1–16, 17–52.

______ . 1972. Altruism and related phenomena, mainly in social insects. *Annu. Rev. Ecol. Syst.* 3:193–232.

Williams, G. C. 1966. *Adaptation and Natural Selection, a Critique of Some Current Evolutionary Thought*. Princeton, N.J.: Princeton University Press, chap. 7.

Trivers, P. L. 1971. The evolution of reciprocal altruism. *Q. Rev. Biol.* 46:35–57.

ALTRUISTIC GENES *see* **ALTRUISTIC TRAITS**

ALTRUISTIC TRAITS [genetics]

= altruistic genes

Haldane, J. B. S. 1932. *The Causes of Evolution*. London: Longmans.

Hamilton, W. D. 1964. The genetical evolution of social behaviour. I and II. *J. Theor. Biol.* 7:1–16, 17–52.

Eshel, I. 1972. On the neighbor effect and the evolution of altruistic traits. *Theor. Pop. Biol.* 3:258–77.

______ . 1977. On the founder effect and the evolution of altruistic traits: an ecogenetical approach. *Theor. Pop. Biol.* 11:410–24.

AMBIGUITY REDUCTION THEORY [genetic code]

Fitch, W. M. 1966. The relation between frequencies of amino acids and ordered trinucleotides. *J. Mol. Biol.* 16:1–8.

______ . 1973. Aspects of molecular evolution. *Annu. Rev.*

Genet. 7:343–80.

Woese, C. 1969. Models for the evolution of codon assignments. *J. Mol. Biol.* 43:235–40.

AMENTIFEROUS CONCEPT [plant taxonomy]

Stern, W. L. 1973. Development of an amentiferous concept. *Brittonia* 25:316–33.

ANABIOSIS [cryobiology]

= SUSPENDED ANIMATION = abiosis = cryptobiosis = viable lifelessness = viability

Leeuwenhoek, A. van. 1702. On certain animalcules found in the sediment in gutters of the roofs of houses. *The Select Works of Anthony van Leeuwenhoek*, Letter 144. Transl. by S. Hoole. 2 vols. London (1798-1807) 2:207–13.

Keilen, D. 1959. The Leeuwenhoek lecture—the problem of anabiosis or latent life: history and current concept. *Proc. Roy. Soc. Lond B* 150:149–91.

Grossowicz, N., S. Hestrin, & A. Keynan, eds. 1961. *Symposium on Cryptobiotic Stages in Biological Systems.* New York: Elsevier.

ANADROMY/CATADROMY PARADOX [fish behavior]

Orr, R. T. 1970. *Animals in Migration.* New York: Macmillan.

Baker, R. R. 1978. *The Evolutionary Ecology of Animal Migration.* New York: Holmes & Meier, chap. 31.

ANAGENESIS [evolution]

Rensch, B. 1947. *Neuere Probleme der Abstammungslehre. Die Transspezifische Evolution.* Stuttgart: Enke.

______. 1959. *Evolution Above the Species Level.* New York: Columbia University Press, pp. 281–308.

Simpson, G. G. 1949. Essay review of recent works on evolutionary theory by Rensch, Zimmermann, and Schindewolf. *Evolution* 3:178–84.

Dobzhansky, T. 1972. Darwinian evolution and the problem of extraterrestrial life. *Perspect. Biol. Med.* 15:157–75.

Cf. CLADOGENESIS

ANALOGOUS VARIATION [evolution]

Darwin, C. 1859. *On the Origin of Species by Means of Natural Selection* . . . . London: Murray, chap. 5.

Meyen, S. V. 1973. Plant morphology in its nomothetical aspects. *Bot. Rev.* 39:205–60.

ANCESTRAL INHERITANCE, LAW OF [genetics]

= law of ancestral heredity = Galton's law

Galton, F. 1865. Hereditary talent and character. *Macmillan's Magazine* 12:157–66, 318–27.

______ . 1897. The average contribution of each several ancestor to the total heritage of the offspring. *Proc. Roy. Soc. Lond.* 61:401–13.

Castle, W. E. 1903. The laws of heredity of Galton and Mendel, and some laws governing race improvement by selection. *Proc. Amer. Acad. Arts Sci.* 39:223–42.

Pearson, K. 1898. Mathematical contributions to the theory of evolution. On the law of ancestral heredity. *Proc. Roy. Soc. Lond.* 62:386–412.

Swinburne, R. G. 1965. Galton's law—formulation and development. *Ann. Sci.* 21:15–31.

Olby, R. C. 1966. *Origins of Mendelism.* New York: Schocken, pp. 68*ff*.

Froggart, P., & N. C. Nevin. 1971. The "law of ancestral heredity" and the Mendelian-ancestrian controversy in England, 1889–1906. *J. Med. Genet.* 8:1–36.

______ . 1971. Galton's "law of ancestral heredity": its influence on the early development of human genetics. *Hist. Sci.* 10:1–27.

ANTECEDENT PLATFORM THEORY [coral reef formation]

Hoffmeister, J. E., & H. S. Ladd. 1944. The antecedent-platform theory. *J. Geol.* 52:387–502.

Stoddart, D. R. 1969. Ecology and morphology of recent coral reefs. *Biol. Rev.* 44:433–98.

ANTHOCORM THEORY [of the flower]

Neumayer, H. 1924. Die Geschichte der Blüte. *Abn. Zool.-Bot. Ges. Wien* 14:1–112.

Meeuse, A. D. J. 1966. *Fundamentals of Phytomorphology.* New York: Ronald.

______ . 1972. Sixty-five years of theories of the multi-axial flower. *Acta Biotheor.* 21:167–202.

ANTHROPOPSYCHIC HYPOTHESIS [of firefly flash synchronization]

Buck, J., & E. Buck. 1966. Biology of synchronous flashing of fireflies. *Nature* 211:562–64.

______ . 1978. Toward a functional interpretation of synchronous flashing by fireflies. *Amer. Nat.* 112:474–92.

ANTIBODY VIRUSES, THEORY OF [immunology]

Smithies, O. 1965. Antibody induction and tolerance. *Science* 149:151–56.

Abramoff, P., & M. F. LaVie. 1970. *Biology of the Immune Response*. New York: McGraw-Hill, p. 253.

ANTIGEN-TEMPLATE THEORY *see* **TEMPLATE HYPOTHESIS**

ANTIGENIC SIMILARITY *see* **MOLECULAR MIMICRY**

ANTI-MONOTONY PRINCIPLE *see* **MONOTONY PRINCIPLE**

ANTITHETIC THEORY [evolution]

Celakovsky, L. 1868. Ubcr die allgemeine Entwickelungsgeschichte des Pflanzenreiches. *Sber. K. Böhm. Ges. Wiss.* 51–67.

———. 1874. Über die verschiedenen Formen und die Bedeutung des Generationswechsels der Pflanzen. *Sber. K. Böhm Ges. Wiss.* 21–61.

Bower, F. O. 1890. On antithetic as distinct from homologous alternation of generations in plants. *Ann. Bot.* 4:347–70.

Lang, W. H. 1909. Discussion on "alternation of generations" at the Linnean Society. *New Phytol.* 8:104–16.

Cf. ALTERNATION OF GENERATIONS; HOMOLOGOUS THEORY

APOSTATIC SELECTION

Clarkc, B. 1962. Balanced polymorphism and the diversity of sympatric species. *Syst. Assoc. Publ.* no. 4, 47–70.

———. 1969. The evidence for apostatic selection. *Heredity* 24:347–52.

Allen, J. A., & B. Clarke. 1968. Evidence for apostatic selection by wild passerines. *Nature* 220:501–02.

Paulson, D. R. 1973. Predator polymorphism and apostatic selection. *Evolution* 27:269–77.

APPARENT COMPETITION [ecology]

Holt, R. D. 1977. Predation, apparent competition, and the structure of prey communities. *Theor. Pop. Biol.* 12:197–229.

APPARENT FREE SPACE [cell physiology]

= free space theory

Hope, A. B., & P. G. Stevens. 1952. Electric potential differences in bean roots and their relation to salt uptake. *Aust. J. Sci. Res. B* 5:335–43.

Lundegardh, H. 1955. Mechanisms of absorption, transport, accumulation, and secretion of ions. *Annu. Rev. Plant Physiol.* 6:1–24.

Briggs, G. E., & R. N. Robertson. 1957. Apparent free space. *Annu. Rev. Plant Physiol.* 8:11–30.

Arisz, W. H. 1961. Symplasm theory of salt uptake into and transport in parenchymatic tissue. *Rec. Adv. Bot.* 2:1125–28.

APPENDICULAR THEORY [of the tubular leaf of *Nepenthes*]

Hooker, J. D. 1859. On the origin and development of the pitchers of *Nepenthes,* with an account of some new Bornean plants of that genus. *Trans. Linn. Soc. Lond.* 22:415–24.

Franck, D. H. 1976. The morphological interpretation of epiascidate leaves. *Bot. Rev.* 42:345–88.

APPENDICULAR THEORY [of the origin of floral structures]

Candolle, A. P. de. 1827. *Organographie Vegetale.* 2 vols. Paris: Germer Bailliere.

Dickinson, T. A. 1978. Epiphylly in angiosperms. *Bot. Rev.* 44:181–232.

ARBOREAL THEORY [bird flight origin]

Marsh, O. C. 1880. Odontornithes: a monograph on the extinct toothed birds of North America. *Engineering Department, U.S. Army, Professional Paper* 18:1–201.

Bock, W. J. 1965. The role of adaptive mechanisms in the origin of higher levels of organization. *Syst. Zool.* 14:272–87.

Ostrom, J. H. 1974. Archaeopteryx and the origin of flight. *Q. Rev. Biol.* 49:27–47.

______ . 1979. Bird flight: how did it begin? *Amer. Sci.* 67:46–56.

ARCHIGENOTYPE *see* **TYPE** [morphological]

ARCHTYPE *see* **TYPE** [morphological]

AREA-CLIMATE HYPOTHESIS [species diversity; biogeography; evolution]

Darlington, P. J. 1957. *Zoogeography: The Geographical Distribution of Animals.* New York: Wiley.

______ . 1959. Area, climate, and evolution. *Evolution* 13:488–510.

Pianka, E. R. 1978. *Evolutionary Ecology*. 2d ed. New York: Wiley, p. 306.

AREA CONCEPT [biogeography]

Rotramel, G. L. 1973. The development of the area concept in biogeography. *Syst. Zool.* 22:227–32.

AREA EFFECTS [on morph frequencies]

Cain, A. J., & J. D. Currey. 1963. Area effects in *Cepaea*. *Phil. Trans. Roy. Soc. Lond. B* 246:1–81.

———. 1963. The causes of area effects. *Heredity* 18:467–71.

Goodhart, C. B. 1963. "Area effects" and non-adaptive variation between populations of *Cepaea* (Mollusca). *Heredity* 18:459–65.

Cf. CLIMATIC SELECTION

AREA-PER SE HYPOTHESIS [species numbers/area]

Preston, F. W. 1960. Time and space and the variation of species. *Ecology* 41:611–27.

Conner, E. F., & E. D. McCoy. 1979. The statistics and biology of the species-area relationship. *Amer. Nat.* 113:791–833.

Cf. SPECIES-AREA CURVE

ARISTOGENESIS [non-Darwinian evolution]

Osborn, H. F. 1934. Aristogenesis, the creative principle in the origin of species. *Amer. Nat.* 68:193–235.

Simpson, G. G. 1953. *The Major Features of Evolution*. New York: Columbia University Press, p. 133.

Cf. ENTELECHY; ORTHOGENESIS

ARRHENIUS' EQUATION [species/area]

Arrhenius, O. 1921. Species and area. *J. Ecol.* 9:95–99.

Gleason, H. A. 1922. On the relation between species and area. *Ecology* 3:158–62.

Goodall, D. W. 1952. Quantitative aspects of plant distribution. *Biol. Rev.* 27:194–245.

Cain, S. A., & G. M. de O. Castro. 1959. *Manual of vegetation analysis*. New York: Harper.

Cf. SPECIES-AREA CURVE

ARRHENIUS' FORMULA [relating biological processes to temperature]

Arrhenius, S. 1915. *Quantitative Laws in Biological Chemistry*. London: Bell, pp. 49*ff*.

ARRESTED EVOLUTION

Ruedemann, R. 1918. The paleontology of arrested evolution.

N.Y. *State Mus. Bull.* no. 196, 107–34.

———. 1922. Additional studies of arrested evolution. *Proc. Natl. Acad. Sci.* 8:54–55.

———. 1922. Further notes on the paleontology of arrested evolution. *Amer. Nat.* 56:256–72.

Simpson, G. G. 1953. *The Major Features of Evolution.* New York: Columbia University Press, chap. 10.

ASPECT DIVERSITY [evolution]

Tinbergen, L. 1960. The natural control of insects in pinewoods. I. Factors influencing the intensity of predation by songbirds. *Arch. Neerl. Zool.* 13:266–336.

Rand, A. S. 1967. Predator-prey interactions and the evolution of aspect diversity. *Atas do Simposio sobre a Biota Amazonica.* 5:73–83.

Ricklefs, R. E., & K. O'Rourke. 1975. Aspect diversity in moths: a temperate-tropical comparison. *Evolution* 29:313–24.

ASSEMBLY RULES [ecology]

Diamond, J. M. 1975. Assembly of species communities. *Ecology and Evolution of Communities.* Edited By M. L. Cody & J. M. Diamond. Cambridge: Harvard University Press, pp. 342–444.

ASSOCIATION [ecology]

Whittaker, R. H. 1951. A criticism of the plant association and climatic climax concepts. *Northwest Sci.* 25:17–31.

Braun, E. L. 1958. The development of association and climax concepts. *Fifty Years of Botany.* Edited by W. C. Steere. New York: McGraw-Hill, pp. 329–39.

ASSOCIATION [wasp nest-building]

Richards, O. W., & M. J. Richards. 1951. Observations on the social wasps of South America *(Hymenoptera Vespidae). Trans. Roy. Entomol. Soc. Lond.* 102:1–170.

Hamilton, W. D. 1964. The genetical evolution of social behaviour. II. *J. Theor. Biol.* 7:17–52.

ASSOCIATION, DOCTRINE OF [plant evolution]

Frost, F. H. 1930. Specialization in secondary xylem of dicotyledons. I. Origin of vessels. *Bot. Gaz.* 89:67–94.

Sporne, K.R. 1956. The phylogenetic classification of the angiosperms. *Biol. Rev.* 31:1–29.

ASSOCIATION-INDUCTION HYPOTHESIS [physiology]

= binding hypothesis

Ling, G. N. 1962. *A Physical Theory of the Living State: The Association-Induction Hypothesis*. New York: Blaisdell.

Palmer, L. G., & J. Gulati. 1976. Potassium accumulation in muscle: a test of the binding hypothesis. *Science* 194:521–23.

ASSOCIATIVE OVERDOMINANCE [genetics]

Frydenberg, O. 1963. Population studies of a lethal mutant in *Drosophila melanogaster*. I. Behaviour in populations with discrete generations. *Hereditas* 50:89–116.

Ohta, T., & M. Kimura. 1971. Behavior of neutral mutants influenced by associated overdominant loci in finite populations. *Genetics* 69:247–60.

Cf. HITCH-HIKING EFFECT

ASSORTIVE MATING [genetics]

Fisher, R. A. 1918. The correlation between relatives on the supposition of Mendelian inheritance. *Trans. Roy. Soc. Edinb.* 52:399–433.

Wright, S. 1921. Systems of mating. III. Assortive mating based on somatic resemblance. *Genetics* 6:143–61.

O'Donald, P. 1960. Assortive mating in a population in which two alleles are segregating. *Heredity* 15:389–96.

Scudo, F. M., & S. Karlin. 1969. Assortative mating based on phenotype. I. Two alleles with dominance. *Genetics* 63:479–98.

ASTRONOMICAL THEORY OF GLACIATION

Zeuner, F. E. 1945. *The Pleistocene Period. Its Climate, Chronology and Faunal Succession*. London: Quaritch.

Good, R. 1974. *The Geography of the Flowering Plants*. 4th ed. London: Longman, p. 423.

ASYMMETRICAL CROSSING-OVER *see* **GENE CONVERSION**

ATAVISM [evolution]

= reversion

Darwin, C. 1883. *The Variation of Animals and Plants Under Domestication*. 2d ed. 2 vols. London: Murray, II, chap. 13.

Arber, A. 1919. On atavism and the law of irreversibility. *Amer. J. Sci.* 48:27–32.

Cf. BIOGENESIS; RECAPITULATION; SPONTANEOUS ATAVISM

ATMUNGSFERMENT CONCEPT [biochemistry]

Warburg, O. 1925. Uber Eisen, den Sauerstoffübertragenden

Bestandteil des Atmungsferments. *Ber. Deut. Chem. Ges.* 48:1001–11.

———. 1925. Iron, the oxygen-carrier of the respiration-ferment. *Science* 61:575–82.

Kohler, R. E. 1973. The background to Otto Warburg's conception of the *Atmungsferment. J. Hist. Biol.* 6:171–92.

ATTENTION STRUCTURE [animal societies]

Chance, M. R. A. 1967. Attention structure as the basis of primate rank orders. *Man* 2:503–18.

———., & C. J. Jolly. 1970. *Social Groups of Monkeys, Apes, and Men*. New York: Dutton.

AUDITORY TEMPLATE HYPOTHESIS [vocal learning in birds]

Konishi, M. 1965. The role of auditory feedback in the control of vocalization in the white-crowned sparrow. *Z. Tierpsychol.* 22:770–83.

———., & F. Nottebohm. 1969. Experimental studies in the ontogeny of avian vocalizations. *Bird Vocalizations*. Edited by R. A. Hinde. London: Cambridge University Press.

Marler, P. A. 1970. A comparative approach to vocal development: song learning in the white-crowned sparrow. *J. Comp. Physiol. Psychol.* 71 (2, pt. 2):1–25.

AUTO-ANTIBODY CONCEPT [immunology]

Tyler, A. 1940. Sperm agglutination in the keyhole limpet, *Megathura crenulata. Biol. Bull.* 78:159–78.

———. 1947. An auto-antibody concept of cell structure, growth and differentiation. *Growth* (Suppl.) 10:7–19.

———. 1955. Ontogeny of immunological properties. *Analysis of Development*. Edited by B. H. Willier, P. A. Weiss, & V. Hamburger. Philadelphia: Saunders, pp. 566–73. (1971 reprint, New York: Hafner.)

AUTOGENESIS *see* **ORTHOGENESIS; NOMOGENESIS; LAMARCKIAN EVOLUTION**

AUTOMATIC SELECTION [evolution of inbreeding]

Fisher, R. A. 1941. Average excess and average effect of a gene substitution. *Ann. Eugen.* 11:53–63.

Jain, S. K. 1976. The evolution of inbreeding in plants. *Annu. Rev. Ecol. Syst.* 7:469–95.

AUTOMIMICRY [feeding behavior]

Brower, L. P. 1970. Plant poisons in the terrestrial food chain and implications for mimicry theory. *Biochemical Evolu-*

tion. Edited by K. L. Chambers. Corvallis: Oregon State University Press.

AUTOTROPH HYPOTHESIS [origin of life]
Hardin, G. 1949. *Biology: Its Human Implications*. San Francisco: Freeman.
Cf. HETEROTROPH HYPOTHESIS; BIOGENESIS

AVOIDANCE [predation]
= non-search image
Clarke, B. 1962. Balanced polymorphism and the diversity of sympatric species. *Syst. Assoc. Publ.* no. 4:47–70.
Cf. SPECIFIC SEARCHING IMAGE

AWARENESS [animal behavior]
Griffin, D. R. 1976. *The Question of Animal Awareness*. New York: Rockefeller University Press.

AXIAL GRADIENT THEORY *see* **GRADIENT THEORY**

AXIAL THEORIES [phylogeny of the megasporangium]
Wardlaw, C. W. 1952. *Phylogeny and Morphogenesis*. London: Macmillan, chap. 19.
Meeuse. A. D. J. 1963. From ovule to ovary: a contribution to the phylogeny of the megasporangium. *Acta Biotheor.* 16:127–82.
Cf. PHYTONIC THEORY

AXILLARY BUD THEORY [of the ovuliferous scale]
Arber, A. 1950. *The Natural Philosophy of Plant Form*. London: Cambridge University Press, pp. 125–31.

[B]

b AND d SELECTION [evolution/population genetics]
Hairston, N. G., D. W. Tinkle, & H. M. Wilbur. 1970. Natural selection and the parameters of population growth. *J. Wildl. Manage.* 34:681–90.
Pianka, F. R. 1972. r and K selection or b and d selection? *Amer. Nat.* 106:581–88.
Cf. r AND K SELECTION

B-RATES *see* **HOROTELY**

 BACTERIAL TRANSFORMATION *see* **TRANSFORMATION** [genetics]

BAER's LAW *see* **VON BAER'S LAWS**

BAKER'S LAW [biogeography]

Baker, H. G. 1955. Self-compatibility and establishment after "long-distance" dispersal. *Evolution* 9:347–49.

______ . 1967. Support for Baker's law—as a rule. *Evolution* 21:853–56.

Stebbins, G. L. 1957. Self fertilization and population variability in the higher plants. *Amer. Nat.* 91:337-54.

Cf. LONG DISTANCE DISPERSAL; REPRODUCTIVE ASSURANCE HYPOTHESIS

BALANCE OF NATURE [ecology]

Hairston, N. G., F. E. Smith, & L. B. Slobodkin. 1960. Community structure, population control, and competition. *Amer. Nat.* 94:421–25.

Murdoch, W. W. 1966. Community structure, population control, and competition—a critique. *Amer. Nat.* 100:219–26.

Ehrlich, P. R., & L. C. Birch. 1967. The "balance of nature" and "population control" *Amer. Nat.* 101:97–107.

Jansen, A. J. 1972. An analysis of "balance in nature" as an ecological concept. *Acta Biotheor.* 21:86–114.

Egerton, F. N. 1973. Changing concepts of the balance of nature. *Q. Rev. Biol.* 48:322–50.

Fretwell, S. D. 1977. The regulation of plant communities by the food chains exploiting them. *Perspect. Biol. Med.* 20:169–85.

BALANCE THEORY [of species differences; evolution]

Carson, H. L. 1975. The genetics of speciation at the diploid level. *Amer. Nat.* 109:83-92.

BALANCE THEORY OF SEX DETERMINATION [genetics]

Goldschmidt, R. 1912. Erblichkeitsstudien an Schmetterlingen. *Z. Indunkt. Abstammungs-Vererbungsl.* 7:1–62.

Stern, C. 1967. Richard Benedict Goldschmidt, 1878–1958. *Biogr. Mem. Natl. Acad. Sci.* 39:141–92.

Allen, G. E. 1974. Opposition to the Mendelian-chromosome theory: the physiological and developmental genetics of Richard Goldschmidt. *J. Hist. Biol.* 7:49–92.

BALANCE-SHIFT THEORY [evolution]

= shifting balance theory

Wright, S. 1931. Evolution in Mendelian populations. *Genetics* 16:97–159.

______. 1932. The roles of mutation, inbreeding, crossbreeding, and selection in evolution. *Proc. 6th Int. Congr. Genet.* 1:356–66.

______. 1956. Modes of selection. *Amer. Nat.* 90:5–24.

______. 1970. Random drift and the shifting balance theory of evolution. *Mathematical Topics in Population Genetics.* Edited by K. Kojima. Berlin: Springer-Verlag, pp. 1–31.

BALANCED ADVANTAGE HYPOTHESIS [of selection]

Ford, E. 1965. *Genetic Polymorphism.* Cambridge: M. I. T. Press.

Johnson, G. B. 1976. Genetic Polymorphisms and enzyme function. *Molecular Evolution.* Edited by F. J. Ayala. Sunderland, Mass.: Sinauer Associates, pp. 46–59.

BALANCED LETHALS, PRINCIPLE OF [genetics]

Muller, H. J. 1918. Genetic variability, twin hybrids and constant hybrids, in a case of balanced lethal factors. *Genetics* 3:422–99.

Carson, H. L. 1967. Permanent heterozygosity. *Evol. Biol.* 1:143–68.

BALANCED MORTALITY HYPOTHESIS [genetic adaptation]

Price, P. W. 1974. Strategies for egg production. *Evolution* 28:76–84.

Cf. BALANCED THEORY

BALANCED POLYMORPHISM *see* **MIMETIC POLYMORPHISM**

BALANCED SELECTION HYPOTHESIS *see* **BALANCED THEORY**

BALANCED THEORY [genetic fitness]

= balance theory = balanced selection hypothesis = balancing selection

Dobzhansky, T. 1955. A review of some fundamental concepts and problems of population genetics. *Cold Spr. Harb. Symp. Quant. Biol.* 20:1–15.

______. 1970. *Genetics of the Evolutionary Process.* New York: Columbia University Press, pp. 126*ff.*

Lewontin, R. C. 1974. *The Genetic Basis of Evolutionary Change.* New York: Columbia University Press, pp. 23*ff.*

Cf. CLASSICAL THEORY; NEUTRAL THEORY

BALANCING SELECTION *see* **BALANCED THEORY**

BALDWIN EFFECT [genetics/evolution]
= ORGANIC SELECTION

Baldwin, J. M. 1896. A new factor in evolution. *Amer. Nat.* 30:441–51, 536-53.

Simpson, G. G. 1953. The "Baldwin effect." *Evolution* 7:110–17.

Waddington, C. H. 1953. The "Baldwin effect," "genetic assimilation" and "homeostasis." *Evolution* 7:386.

Mayr, E. 1963. *Animal Species and Evolution*. Cambridge: Belknap Press of Harvard University Press, pp. 610–12.

BALFOUR'S LAW [of cell division]

Balfour, F. M. 1875. A comparison of the early stages in the development of vertebrates. *Q. J. Microsc. Sci.* 15:207–26.

Løvtrup, S. 1974. *Epigenetics*. New York: Wiley, pp. 128–29.

Rappaport, R. 1974. Cleavage. *Concepts of Development*. Edited by J. Lash & J. R. Whittaker. Stamford, Conn.: Sinauer Associates, pp. 76-98.

BANCROFT'S LAW [adjustment or response to strain]

Bancroft, W. D. 1911. A universal law. *Science*, n.s., 33:159–79.

Adams, C. C. 1913. *Guide to the Study of Animal Ecology*. New York: Macmillan, chap. 7.

BATEMAN'S PRINCIPLE [of reproductive success; sexual selection]

Bateman, A. J. 1948. Intra-sexual selection in *Drosophila*. *Heredity* 2:349–68.

Cf. PARENTAL INVESTMENT, THEORY OF

BATESIAN MIMICRY [evolution]
= Bates theory of mimicry

Bates, H. W. 1862. Contributions to an insect fauna of the Amazon valley. *Lepidoptera: Heliconidae. Trans. Linn. Soc. Lond.* 23:495–565.

Carpenter, G. D. H. & E. B. Ford. 1933. *Mimicry*. London: Methuen.

Sheppard, P. M. 1959. The evolution of mimicry: a problem in ecology and genetics. *Cold Spr. Harb. Symp. Quant. Biol.* 24:131–40.

Evans, M. A. 1965. Mimicry and the Darwinian heritage. *J. Hist. Ideas* 26:211–20.

Wickler, W. 1968. *Mimicry in Plants and Animals*. New York: McGraw-Hill, chap. 1.

Beddall, B. G., ed. 1969. *Wallace and Bates in the Tropics.* London: Macmillan, pp. 10, 211*ff*.

Nur, O. 1970. Evolutionary rates of models and mimics in Batesian mimicry. *Amer. Nat.* 104:477–86.

Sternburg, J. G., G. P. Waldbauer, & M. R. Jeffords. 1977. Batesian mimicry: selective advantage of color pattern. *Science* 195:681–83.

Cf. MIMICRY

BEACON HYPOTHESIS [of firefly flashing synchronization]

Buck, J., & E. Buck. 1968. Mechanism of rythmic synchronous flashing of fireflies. *Science* 159:1319–27.

______ . 1978. Toward a functional interpretation of synchronous flashing by fireflies. *Amer. Nat.* 112:471-92.

BEANBAG GENETICS

Mayr, E. 1959. Where are we? *Cold Spr. Harb. Symp. Quant. Biol.* 24:1–14.

Wright, S. 1960. "Genetics and Twentieth Century Darwinism." A review and discussion. *Amer. J. Hum. Genet.* 12:365–72.

Haldane, J. B. S. 1964. A defense of beanbag genetics. *Perspect. Biol. Med.* 7:343–59.

BEAT THEORY [of bat echo-location]

Pye, J. D. 1960. A theory of echolocation by bats. *J. Laryngol. Otol.* 74:718–29.

______ . 1961. Echolocation by bats. *Endeavour* 20:101–11.

______ . 1963. Mechanisms of echolocation. *Ergeb. Biol.* 26:12–20.

Kay, L. 1962. A plausible explanation for bats' echo-location acuity. *Anim. Behav.* 10:34–41.

Fulton, M. B. 1974. The role of echolocation in the evolution of bats. *Amer. Nat.* 108:386–88.

BEAU GESTE HYPOTHESIS [bird behavior]

Krebs, J. R. 1977. The significance of song repertoires: the beau geste hypothesis. *Anim. Behav.* 25:475–78.

BEE LANGUAGE CONTROVERSY [bee behavior]

Wenner, A. M. 1971. *The Bee Language Controversy*. Boulder, Col.: Educational Programs Improvement.

Wells, P. H., & A. M. Wenner. 1973. Do honey bees have a language? *Nature* 241:171–75.

Frisch, K. von. 1973. Review of "the bee language controversy." *Anim. Behav.* 21:628-30.

Ferguson, A. 1975. Evolution, von Frisch, and teleology. *Amer. Nat.* 109:369–70.

BEHAVIORAL HOMEOSTASIS *see* **DEVELOPMENTAL HOMEOSTASIS**

BENEFICIAL DEATH [ecology]

Allee, W. C., A. E. Emerson, O. Park, T. Park, & K. P. Schmidt. 1949. *Principles of Animal Ecology.* Philadelphia: Saunders, pp. 692*ff.*

Emerson, A. F. 1960. The evolution of adaptation in population systems. *The Evolution of Life.* Edited by S. Tax. Chicago: University of Chicago Press, pp. 307–48.

Bonner, J. T. 1965. *Size and Cycle.* Princeton, N.J.: Princeton University Press, p. 170.

BERGMANN-NIEMANN THEORY [amino-acid sequence in proteins]

Bergmann, M., & C. Niemann. 1936. On blood fibrin; a contribution to the problem of protein structure. *J. Biol. Chem.* 115:77–85.

Astbury, W. T. 1943. X-rays and the stoichiometry of the proteins. *Adv. Enzymol.* 3:63–108.

Needham, A. E. 1964. *The Growth Process in Animals.* London: Pitman, p. 201.

BERGMANN'S RULE [relating climate and faunistic distribution]

Bergmann, C. 1847. Über die Verhältnisse der Wärmeökonomie der Thiere zu iher Grösse. *Göttinger Studien* 3, pt. 1:595–708.

Scholander, P. F. 1955. Evolution of climatic adaptation in homeotherms. *Evolution* 9:15–26.

______ . 1956. Climatic rules. *Evolution* 10:339–40.

Mayr, E. 1956. Geographical character gradients and climatic adaptation. *Evolution* 10:105–08.

Williamson, K. 1958. Bergmann's rule and obligatory overseas migration. *Brit. Birds* 51:209–31.

Ray, C. 1960. The application of Bergmann's and Allen's rules to the poikilotherms. *J. Morph.* 105:85–108.

Kendiegh, S. C. 1969. Tolerance of cold and Bergmann's rule. *Auk* 86:13–25.

James, F. C. 1970. Geographic size variation in birds and its relationship to birds. *Ecology* 51:365–90.

McNab, B. K. 1971. On the ecological significance of Bergmann's rule. *Ecology* 52:845–54.

Whitford, P. C. 1979. An explanation of altitudinal deviations from Bergmann's rule as applied to birds. *Biologist* 61:1–10.

Cf. NEO-BERGMANNIAN RULE

BERNARD'S PRINCIPLE [physiology]

= milieu intérieur

Les Concepts de Claude Bernard sur le milieu intérieur . . . . 1967. Paris: Masson.

Holmes, F. L. 1962. Claude Bernard and the concept of the internal environment. Ph.D. dissertation, Harvard University.

———. 1963. Claude Bernard and the milieu intérieur. *Arch. Int. Hist. Sci.* 16:369–76.

———. 1969. Joseph Barcroft and the fixity of the internal environment. *J. Hist. Biol.* 2:89–122.

Langley, L. L., ed. 1973. *Homeostasis: Origins of the Concept.* Stroudsburg, Pa.: Dowden Hutchinson & Ross, section II.

BERNSTEIN MEMBRANE THEORY [of electrical transients of excitable tissues]

Bernstein, J. 1902. Untersuchungen zur Thermodynamik der bioelektrischen Ströme. *Pflügers Arch. Ges. Physiol.* 92:521–62.

———. 1912. *Elektrobiologie.* Braunschweig: Fr. Vieweg.

Hodgkin, A. L. 1964. *The Conduction of the Nervous Impulse.* Springfield, Ill.: Thomas, chap. 3.

Noble, D. 1966. Applications of Hodgkin-Huxley equations to excitable tissue. *Physiol. Rev.* 46:1–50.

Hille, B. 1970. Ionic channels in nerve membranes. *Prog. Biophys. Mol. Biol.* 21:1–32.

Cf. IONIC THEORY

BERTALANFFY'S THEORY [growth, metabolism]

= Bertalanffy's growth equations

Bertalanffy, L. von. 1942. *Theoretische Biologie.* vol. II: *Stoffwechsel, Wachstum.* Berlin: Gebrüder Borntraeger.

———. 1950. The theory of open systems in physics and biology. *Science* 111:23–29.

———. 1951. Metabolic types and growth types. *Amer. Nat.* 85:111–17.

Nowinski, W. W. 1960. *Fundamental Aspects of Normal and Malignant Growth.* Amsterdam: Elsevier, pp. 193*ff.*

BI-COMPONENT HYPOTHESIS [compass orientation in insects]

Mittelstaedt, H. 1962. Control systems of orientation in insects. *Annu. Rev. Entomol.* 7:177–98.

BIDIRECTIONAL TRANSPORT CONCEPT [plant physiology]

Zimmermann, M. H. 1974. Long distance transport *Plant Physiol.* 54:472–79.

______ . , & J. A. Milburn, eds. 1975. *Transport in Plants.* vol I: Phloem Transport. Berlin: Springer-Verlag, chap. 10.

BIFACIALITY THEORY [of the tubular leaf of Nepenthes]

Roth, I. 1953. Zur entwicklungsgeschichte und histogenese der Schlauchblätter von *Nepenthes. Planta* 42:177–208.

Franck, D. H. 1976. The morphological interpretation of epiascidate leaves. *Bot. Rev.* 42:345–88.

BIG-BANG REPRODUCTION STRATEGY [single, suicidal breeding effort; ecology]

Gadgil, M., & W. H. Bossert. 1970. Life historical consequences of natural selection. *Amer. Nat.* 104:-1–24.

Schaffer, W. M., & M. D. Gadgil. 1975. Selection for optimal life histories in plants. *Ecology and Evolution of Communities.* Edited by M. L. Cody & J. M. Diamond. Cambridge: Harvard University Press, pp. 142–57.

BIG MOTHER HYPOTHESIS [reproductive adaptation]

Ralls, K. 1976. Mammals in which females are larger than males. *Q. Rev. Biol.* 51:245–76.

Myers, P. 1978. Sexual dimorphism in size of vespertilionid bats. *Amer. Nat.* 112:701–11.

BILATEROGASTREA THEORY [origin of Metozoa]

Jägerston, G. 1955. On the early phylogeny of the Metazoa. The bilaterogastrea theory. *Zool. Bidr. Upps.* 30:321–54.

Cf. GASTREA THEORY

BINDING HYPOTHESIS *see* **ASSOCIATION-INDUCTION HYPOTHESIS**

BIOCHROMATICAL LAWS [evolution]

Peterich, L. 1972. Biological chromatology. The laws of colour and design in nature. *Acta Biotheor.* 21:24–46.

BIOCLIMATIC LAW [biogeography]

= Hopkins' bioclimatic law

Hopkins, A. D. 1918. Periodic events and natural law as guides to agricultural research and practice. *U.S.D.A. Monthly Weather Review,* suppl. no. 9.

______ . 1921. Intercontinental problems in bioclimatics. *J. Wash. Acad. Sci.* 11:223–27.

——— . 1938. Bioclimatics. *U.S.D.A. Misc. Publ.* no. 280.
Shelford, V. E. 1929. *Laboratory and Field Ecology*. Baltimore: Williams & Wilkins, chap. 1.
Hardwick, D. F. 1971. The "phenological date" as an indicator of the flight period of Noctuid moths. *Can. Entomol.* 103:1207–16.

BIOENERGETICS

Szent-Györgyi, A. 1957. *Bioenergetics*. New York: Academic Press.
Lehninger, A. 1965. *Bioenergetics*. New York: Benjamin.
Peusner, L. 1974. *Concepts in Bioenergetics*. Englewood Cliffs, N.J.: Prentice-Hall.

BIOGENESIS ["ontogeny recapitulates phylogeny"]
= biogenetic law = RECAPITULATION

Haeckel, E. 1866. *Generelle Morphologie der Organismen*. 2 vols. Berlin: Reimer.
——— . 1874. Die Gastraeatheorie, die phylogenetische Classifikation des Tierreichs und die Homologie der Keimblätter. *Jena. Z. Naturw.* 8:1-55.
——— . 1900. *The Riddle of the Universe at the Close of the Nineteenth Century*. New York: Harper, p. 81.
Garstang, W. 1922. The theory of recapitulation: a critical restatement of the biogenetic law. *J. Linn. Soc. Lond. Zool.* 35:81–101.
Radl, E. 1930. *The History of Biological Theories*. Transl. by E. J. Hatfield. London: Oxford University Press, chap. 12.
Shumway, W. 1932. The recapitulation theory. *Q. Rev. Biol.* 7:93–99.
de Beer, G. R. 1951. *Embryos and Ancestors*. Rev. ed. Oxford: Clarendon Press, p. 7*ff*.
Oppenheimer, J. M. 1955. Problems, concepts and their history. *Analysis of Development*. Edited by B. H. Willier, P. A. Weiss, & V. Hamburger. New York: Saunders, pp. 1–23.
Smit, P. 1962. Ontogenesis and phylogenesis: their interrelation and their interpretation. *Acta Biotheor.* 15:1–104.
Doetsch, R. N. 1963. Studies on biogenesis by Sir William Roberts. *Med. Hist.* 7:232–40.
Gould, S. J. 1977. *Ontogeny and Phylogeny*. Cambridge: Harvard University Press.
Maienschein, J. 1978. Cell lineage, ancestral reminiscence, and the biogenetic law. *J. Hist. Biol.* 11:129–58.
Nelson, G. 1978. Ontogeny, phylogeny, paleontology, and the biogenetic law. *Syst. Zool.* 27:324–45.

Gould, S. J. 1979. On the importance of heterochrony for evolutionary biology. *Syst. Zool.* 28:224–26.

Cf. VON BAER'S LAWS; ATAVISM

BIOGEOGRAPHIC THEORY

Hofsten, N. von. 1916. Zur älteren Geschichte des Diskontinuitäts problems in der Biogeographie. *Zool. Ann.* 7:197–353.

Egerton, F. N. 1968. Studies of animal populations from Lamarck to Darwin. *J. Hist. Biol.* 1:225–59.

Cf. EQUILIBRIUM THEORY OF ISLAND BIOGEOGRAPHY; RECAPITULATION

BIOLOGICAL CLOCK [physiology]

= physiological clock

Cold Spring Harbor Symposium on Quantitative Biology. 1960. vol. 25: *Biological Clocks.*

Moore, S. 1967. *Biological Clocks and Patterns.* New York: Criterion.

Brown, F. A. 1970. *The Biological Clock.* New York: Academic Press.

Laar, W. van. 1970. A contribution to the problem of the concept "biological clock." (Part I) *Acta Biotheor.* 19:95–139.

Bünning, E. 1973. *The Physiological Clock.* 3d Engl. ed. New York: Academic Press.

Hastings, J. W. 1972. Timing mechanisms. *Challenging Biological Problems.* Edited by J. A. Behnke. New York: Oxford University Press, pp. 148–67.

Palmer, J. D. 1973. Tidal rhythms: the clock control of the rhythmic physiology of marine organisms. *Biol. Rev.* 48:377–418.

______ . 1974. *Biological Clocks in Marine Organisms.* New York: Wiley.

Cf. CIRCADIAN RHYTHM

BIOLOGICAL COMPENSATION

Conrad, M. 1976. Biological adaptability: the statistical state model. *Bioscience* 26:319–24.

BIOLOGICAL CONTROL [of animal populations] *see* **NATURAL CONTROL**

BIOLOGICAL EFFICIENCY *see* **ECOLOGICAL EFFICIENCY**

BIOLOGICAL EXPLOITATION THEORY

Rosenzweig, M. L. 1977. Aspects of biological exploitation. *Q. Rev. Biol.* 52:371–80.

Cf. PREDATION HYPOTHESIS

BIOLOGICAL MAINTENANCE, LAW OF

Bertalanffy, L. von. 1933. *Modern Theories of Development.* Transl. by J. H. Woodger. London: Oxford University Press.

Arber, A. 1941. The interpretation of leaf and root in the angiosperms. *Biol. Rev.* 16:81–105.

Cf. PHYTON THEORY

BIOLOGICAL MEMORY

Walker, I. 1972. Biological memory. *Acta Biotheor.* 21:203–35.

BIOLOGICAL MOMENTUM, LAW OF

= Abel's law

Abel, O. 1929. *Paläobiologie und Stammegeschichte.* Jena: Fischer.

Hennig, W. 1966. *Phylogenetic Systematics.* Urbana: University of Illinois Press, p. 220.

BIOLOGICAL SPECIES

Löve, A. 1964. The biological species concept and its evolutionary structure. *Taxon* 13:33–44.

Mayr, E. 1969. The biological meaning of species. *Biol. J. Linn. Soc.* 1:311–20.

Sokal, R. R., & T. J. Crowells. 1970. The biological species concept: a critical evaluation. *Amer. Nat.* 104:127–53.

Cf. SPECIES CONCEPTS

BIOLOGICAL THEORY OF EVOLUTION *see* **SYNTHETIC THEORY OF EVOLUTION**

BIOME [ecology]

Tansley, A. G. 1935. Use and abuse of vegetational concepts and terms. *Ecology* 16:284–307.

Shelford, V. E. 1945. The relative merits of the life zone and biome concepts. *Wilson Bull.* 57:248–52.

Whittaker, R. H. 1975. *Communities and Ecosystems.* 2d ed. New York: Macmillan, pp. 135–61.

BIOPOLARITY [in discontinuous distribution of animal species]

Pfeffer, G. 1891. *Versuch über die erdgeschichtliche Entwicklung der jetzigen Verbreitungsverhältnisse unserer Tierwelt.* Hamburg: Freiderichsen, p. 38.

Ortmann, A. E. 1899. On new facts lately presented in opposition to the hypothesis of bipolarity of marine fauna. *Amer. Nat.* 33:583–91.

Allee, W. C., & K. P. Schmidt. 1951. *Ecological Animal Geography*. 2d ed., rev. by R. Hesse. New York: Wiley, pp. 333*ff*.

BIOSPACE [ecology]

Doty, M. S. 1957. Rocky intertidal surfaces. *Treatise on Marine Ecology and Paleoecology*. Edited by J. W. Hedgpeth. *Geol. Soc. Amer. Mem.* 67:I:535–85.

Valentine, J. W. 1969. Patterns of taxonomic and ecological structure of the shelf benthos during Phanerozoic time. *Palaeontology* 12:684-709.

Cf. ECOSPACE

BIOTIC COMMUNITY [ecology]

Shelford, V. E. 1912. Ecological succession. IV. Vegetation and the control of land animal communities. *Biol. Bull.* 23:59–99.

Vestal, A. G. 1914. Internal relations of terrestrial associations. *Amer. Nat.* 48:413–45.

Phillips, J. 1931. The biotic community. *J. Ecol.* 19: 1–24.

Clements, F. E., & V. E. Shelford. 1939. *Bio-ecology*. New York: Wiley, p. 5.

BIOTIC POTENTIAL [ecology]

= intrinsic rate of natural increase

Chapman, R. N. 1928. The quantitative analysis of environmental factors. *Ecology* 9:111–22.

Cole, L. C. 1954. The population consequences of life history phenomena. *Q. Rev. Biol.* 29:103–37.

BIOTROPHY [Parasitism]

Thrower, L. B. 1966. Terminology for plant parasites. *Phytopathol. Zeit.* 52:319–34.

Lewis, D. H. 1973. Concepts in fungal nutrition and the origin of biotrophy. *Biol. Rev.* 45:261–78.

______ . 1974. Micro-organisms and plants: the evolution of parasitism and mutualism. *Symp. Soc. Gen. Microbiol.* 24:367–92.

Cf. MUTUALISM; SYMBIOSIS

BLENDING INHERITANCE

= blending concept of heredity = blood theory of heredity = PAINT-POT THEORY OF HEREDITY

Fisher, R. A. 1930. *The Genetical Theory of Natural Selection.* Oxford: Clarendon Press, chap. 1.

Dobzhansky, T. 1951. *Genetics and the Origin of Species.* 3d ed., rev. New York: Columbia University Press, pp. 51–52.

———. F. J. Ayala, G. L. Stebbins, & J. W. Valentine. 1977. *Evolution.* San Francisco: Freeman, chap. 4.

Vorzimmer, P. J. 1963. Charles Darwin and blending inheritance. *Isis* 54:371–90.

BODY-CAP CONCEPT [of the apical meristem]

Schüepp, O. 1917. Untersuchungen über Wachstum und Formwechsel von Vegetationspunkten. *Jahrb. Wiss. Bot.* 57:17–79.

Esau, K. 1977. *Anatomy of Seed Plants.* 2d ed. New York: Wiley, p. 228.

BODYGUARD HYPOTHESIS [genetics]

Hsu, T. C. 1975. A possible function of constitutive heterochromatin: the bodyguard hypothesis. *Genetics* 79 (Suppl.):137–50.

BOIVIN-VENDRELY RULE [of DNA content in cells]

Boivin, A., R. Vendrely, & C.Vendrely. 1948. L'acide désoxyribonucléique du noyau cellulaire, dépositaire des caractères héréditaires; arguments d'ordre analytique. *Acad. Sci.* 226:1061–63.

Mirsky, A. E., & H. Ris. 1949. Variable and constant components of chromosomes. *Nature* 163:666–67.

Olby, R. 1974. DNA before Watson-Crick. *Nature* 248:782–85.

BOLK'S CONCENTRATION THEORY *see* **CONCENTRATION THEORY**

BOLK's FETALIZATION THEORY *see* **FETALIZATION THEORY**

BORELLI'S LAW [animal physiology]

Borelli, G. A. 1680. *De Motu Animalium.* Rome: Bernaho.

Thompson, D'A. W. 1942. *On Growth and Form.* 2d ed. Cambridge: Cambridge University Press, pp. 36-39.

BOTTLENECK EFFECT

= eye of a needle

Nei, M., T. Maruyama, & R. Chakraborty. 1975. The bottleneck effect and genetic variability in populations. *Evolution* 29:1–10.

Dodson, E. O., & P. Dodson. 1976. *Evolution: Process and Product*. 2d ed. New York: Van Nostrand, p. 334.

Soule, M. 1976. Allozyme variation: its determinants in space and time. *Molecular Evolution*. Edited by F. J. Ayala. Sunderland, Mass.: Sinauer Associates, pp. 60–77.

Chakraborty, R., & M. Nei. 1977. Bottleneck effects on average heterozygosity and genetic distance with the stepwise mutation model. *Evolution* 31:347–56.

Cf. FOUNDER PRINCIPLE

BRADYTELY [rate distribution in evolution]

Simpson, G. G. 1944. *Tempo and Mode in Evolution*. New York: Columbia University Press, chap. 4.

______ . 1953. *The Major Features of Evolution*. New York: Columbia University Press, chap. 10.

Cf. HOROTELY; TACHYTELY

BREAKAGE-AND-COPYING HYPOTHESIS [genetic recombination]

Delbrück, M., & G.S. Stent. 1957. On the mechanism of DNA replication. *The Chemical Basis of Heredity*. Edited by E. D. McElroy & B. Glass. Baltimore: Johns Hopkins Press, pp. 699–736.

Boon, T., & N. D. Zinder. 1969. A mechanism for genetic recombination generating one parent and one recombinant. *Proc. Natl. Acad. Sci.* 64:573–77.

______ . 1971. Genotypes produced by individual recombination events involving bacteriophage f1. *J. Mol. Biol.* 58:133–51.

BREAKAGE-REUNION HYPOTHESIS [chromosome mutation]

Stadler, L. J. 1932. On the genetic nature of induced mutations in plants. *Proc. 6th Int. Congr. Genet.* 1:274–94.

Evans, H. J. 1962. Chromosome aberrations induced by ionizing radiation. *Int. Rev. Cytol.* 13:221–321.

BRETSKY AND LORENZ THEORY [community diversity]

Bretsky, P. W., & D. M. Lorenz. 1970. Adaptive response to environmental stability: a unifying concept in paleoecology. *Proceedings of the North American Paleontological Convention*, Chicago, September, 1969. Lawrence, Kan.: Allen Press, Pt. E, pp. 522–50.

Thayer, C. W. 1973. Taxonomic and environmental stability in the Paleozoic. *Science* 182:1242–43.

Valentine, J. W., & F. J. Ayala. 1974. On scientific hypotheses, killer clams, and extinctions. *Geology* 2:69–71.

Gould, S. J. 1976. Palaeontology plus ecology as palaeobiology. *Theoretical Ecology: Principles and Applications*. Edited by R. M. May. Oxford: Blackwell, pp. 218–36.

BRITTEN-DAVIDSON MODEL [of gene regulation]

Britten, R. J., & E. H. Davidson. 1969. Gene regulation for higher cells: a theory. *Science* 165:349–57.

———. 1971. Repetitive and non-repetitive DNA sequences and a speculation on the origins of evolutionary novelty. *Q. Rev. Biol.* 46:111–38.

Davidson, E. H., & R. J. Britten. 1973. Organization, transcription, and regulation in the animal genome. *Q. Rev. Biol.* 48:565–613.

BROKEN-STICK MODEL [species abundance; ecology]

= MacArthur's broken stick model = MacArthur model

MacArthur, R. H. 1957. On the relative abundance of bird species. *Proc. Natl. Acad. Sci.* 43:293–95.

———. 1960. On the relative abundance of species. *Amer. Nat.* 94:25–36.

King, D. E. 1964. Relative abundance of species and MacArthur's model. *Ecology* 45:716–27.

DeVita, J. 1979. Niche separation and the broken-stick model. *Amer. Nat.* 114:171-78.

BROOKS'S LAW *see* **DYAR'S RULE**

BROWNIAN MOTION [biophysics; cell biology]

Brown, R. 1866. *The Miscellaneous Botanical Works of Robert Brown*. London: Ray Society, I:463–86.

Barnes, R. B., & S. Silverman. 1934. Brownian motion as a natural limit to all measuring processes. *Rev. Mod. Phys.* 6:162–92.

Wang, N. C., & G.E. Uhlenbeck. 1945. On the theory of the Brownian motion. *Rev. Mod. Phys.* 7:323–42.

Einstein, A. 1926 *Investigations on the Theory of the Brownian Movement*. London: Methuen (repr. Dover, 1956).

Kerker, M. 1976. The Svedberg and molecular reality. *Isis* 67:190–216.

BRUCE EFFECT [animal physiology]

= natural birth control

Bruce, H. M. 1960. A block to pregnancy in the mouse caused by proximity of strange males. *J. Reprod. Fertil.* 1:96–103.

———. 1966. Smell as an exteroceptive factor. *J. Anim. Sci.* 25 (Suppl.):83–89.

Schwagmeyer, P.L. 1979. The Bruce effect: an evaluation of male/female advantages. *Amer. Nat.* 114:932–38.

BUFFER EFFECT [population ecology]

Kluyner, H. N., & L. Tinbergen. 1953. Territory and regulation of density in titmice. *Arch. Neerl. Zool.* 10:265–87.

Brown, J. L. 1969. The buffer effect and productivity in tit populations. *Amer. Nat.* 103:347–54.

BUFFON'S LAW [species/area]

Nelson, G. 1978. From Candolle to Croizat: comments on the history of biogeography. *J. Hist. Biol.* 11:269–305.

Cf. ALLOPATRIC SPECIATION

BULLER PHENOMENON

Buller, A. H. R. 1930. The biological significance of conjugate nuclei in *Coprinus lagopus* and other Hymenomycetes. *Nature* 126:686–89.

Quintanilha, A. 1937. Contribution a l'étude genetique du phenomene de Buller. *C.R. Acad.Sci.* 205:745–47.

Raper, J. R. 1966. *Genetics of Sexuality in Higher Fungi.* New York: Ronald, chap. 8.

BUNNING'S CONCEPT [plant organization]

Bünning, E. 1948. *Entwicklungs- und Bewegungs-physiologie der Pflanzen.* Berlin: Springer-Verlag.

______ . 1952. Morphogenesis in plants. *Surv. Biol. Prog.* 2:105–40.

Wardlaw, C. W. 1965. *Organization and Evolution in Plants.* London: Longmans, p. 162.

[C]

C-RATES [of evolution]

Kurtén, B. 1959. Rates of evolution in fossil mammals. *Cold Spr. Harb. Symp. Quant. Biol.* 24:205–15.

CABLE THEORY [of cell membrane resistance]

Hodgkin, A. L., & W. A. H. Rushton. 1946. The electrical constants of a crustacean nerve fibre. *Proc. Roy. Soc. Lond. B* 133:444–79.

Taylor, R. E. 1963. Cable theory. *Physical Techniques in*

Biological Research. Vol. 6B. Edited by W. L. Nastuk. New York: Academic Press, pp. 219–62.

Noble, D. 1966. Applications of Hodgkin-Huxley equations to excitable tissue. *Physiol. Rev.* 46:1–50.

CAENOGENESIS [evolution]

Haeckel, E. 1866. *Morphologie der Organismen*. 2 vols. Berlin: Reimer, vol. 2.

de Beer, G. R. 1958. *Embryos and Ancestors*. 3d ed. Oxford: Clarendon Press, chaps. 1, 6.

Cf. BIOGENESIS

CANALIZATION [genetics]

= developmental canalization

Waddington, C. H. 1942. Canalization of development and the inheritance on acquired characters. *Nature* 150:563–65.

______ . 1957. *The Strategy of the Genes*. London: Allen & Unwin, chap. 2.

Rendel, J. M. 1967. *Canalisation and Gene Control*. London: Logos.

CANONICAL DISTRIBUTION [species abundance]

Preston, F. W. 1962. The canonical distribution of commonness and rarity. I and II. *Ecology* 43:185–214, 410–32.

May, R. M. 1975. Patterns of species abundance and diversity. *Ecology and Evolution of Communities*. Edited by M. L. Cody & J. M. Diamond. Cambridge: Harvard University Press, pp. 81-120.

Cf. RED QUEEN HYPOTHESIS; SPECIES-AREA CURVE

CARPEL [plant morphology]

Lorch, J. 1963. The carpel—a case-history of an idea and a term. *Centaurus* 8:269–91.

Cf. METAMORPHOSIS, THEORY OF

CARPEL POLYMORPHISM, THEORY OF [plant morphology]

Saunders, E. R. 1925, 1927, 1929. On carpel polymorphism. I. II. III. *Ann. Bot.* 39:123–67; 41:569–627; 43:459–81.

Eames, A. J. 1931. The vascular anatomy of the flower with refutation of the theory of carpel polymorphism. *Amer. J. Bot.* 18:147–88.

CARRIER CONCEPT [ion transport]

Patlak, C. S. 1956. Contributions to the theory of active transport. *Bull. Math. Biophys.* 18:271–313.

Overstreet, R. 1961. On the nature of ion carriers in mineral absorption by plants. *Recent Advances in Botany*. 2 vols. Toronto: University of Toronto Press. 2:1111–16.

Rehm, W. S. 1964. Hydrochloric acid secretion, ion gradients, and the gastric potential. *The Cellular Functions of Membrane Transport*. Edited by J. F. Hoffman. Englewood Cliffs, N.J.: Prentice-Hall, pp. 231–50.

Kotyk, A., & K. Janáček. 1975. *Cell Membrane Transport*. 2d ed. New York: Plenum, chap. 8.

CARRYING CAPACITY [ecology]

Revelle, R. 1974. Food and population. *Sci. Amer.* 231:161–70.

Brown, L. 1978. *The Twenty-ninth Day: Accommodating Human Needs and Numbers to the Earth's Resources*. New York: Norton, pp. 13*ff*.

CARTESIAN TRANSFORMATIONS *see* **ORTHOGENESIS**

CASTE POLYETHISM *see* **POLYETHISM**

CASTLE-HARDY-WEINBERG PRINCIPLE/LAW *see* **HARDY-WEINBERG LAW; SQUARE LAW**

CASTLE'S LAW [genetic equilibrium]
= HARDY-WEINBERG LAW

Castle, W. E. 1903. The laws of Galton and Mendel and some laws governing race improvement by selection. *Proc. Amer. Acad. Arts Sci.* 35:233–42.

Li, C. C. 1967. Castle's early work on selection and equilibrium. *Amer. J. Hum. Genet.* 19:70–74.

Keeler, C. 1968. Some oddities in the delayed appreciation of "Castle's law." *J. Hered.* 59:110–12.

CASTLE'S RAT TRAP THEORY [of linear arrangement of genes]

Castle, W. 1919. Is the arrangement of the genes in the chromosome linear? *Proc. Natl. Acad. Sci.* 5:25–32.

Muller, H. J. 1920. Are the factors of heredity arranged in a line? *Amer. Nat.* 54:97–121.

Carlson, E. A. 1966. *The Gene: A Critical History*. Philadelphia: Saunders, pp. 79–81.

CATASTROPHE THEORY [evolution]

Thom, R. 1976. *Structural Stability and Morphogenesis*. Transl. by A. D. Fowler. Reading, Mass.: Benjamin.

Zeeman, E. C. 1976. Catastrophe theory. *Sci. Amer.* 234(4):65–83.

Kolata, G. B. 1977. Catastrophe theory: the emperor has no clothes. *Science* 196:287, 350–51.

Zahler, R. S., & H. J. Sussmann. 1977. Claims and accomplishments of applied catastrophe theory. *Nature* 269:759–63.

Cf. THOM'S THEORY OF CATASTROPHIES; FOLD CATASTROPHE

CATASTROPHIC MORTALITY *see* **CATASTROPHIC SELECTION**

CATASTROPHIC SELECTION

= catastrophic mortality

Lewis, H. 1962. Catastrophic selection as a factor in speciation. *Evolution* 16:257–71.

CATASTROPHIC SPECIATION

Lewis, H. 1962. Catastrophic selection as a factor in speciation. *Evolution* 16:257–71.

——— . 1966. Speciation in flowering plants. *Science* 152:167–72.

Dobzhansky, T., F. J. Ayala, G. L. Stebbins, & J. W. Valentine. 1977. *Evolution*. San Francisco: Freeman, chap. 7.

CATASTROPHISM [evolution]

Hooykaas, R. 1957. The parallel between the history of the earth and the history of the animal world. *Arch. Int. Hist. Sci.* 38:3–18.

Cannon, W. F. 1960. The uniformitarian-catastrophist debate. *Isis* 51:38–55.

Brooks, J. L. 1972. Extinction and the origin of organic diversity. *Trans. Conn. Acad. Arts Sci.* 44:17–56.

Cf. PHYLETIC GRADUALISM; PROGRESSIONISM; UNIFORMITARIANISM

CELL AS UNIT THEORY [cell physiology]

Weiss, P. 1962. From cell to molecule. *The Molecular Control of Cellular Activity*. Edited by J. M. Allen. New York: McGraw-Hill, pp. 1–72.

——— . 1963. The cell as unit. *J. Theor. Biol.* 5:389–97.

CELL CHAIN THEORY [of nerve development]

Schwann, T. 1838/9. *Mikroskopische Untersuchungen*, ... Berlin. Reprinted in *Ostwald's Klassiker der exakten Wissenschaften*, no. 176, Leipzig, Engelmann, 1910.

Detwiler, S. R. 1933. Experimental studies upon the development of the amphibian nervous system. *Biol. Rev.* 8:269–310.

Harrison, R. G. 1969. *Organization and Development of the Embryo.* New Haven, Conn.: Yale University Press, p. 122.

CELL DIFFERENTIATION

Weismann, A. 1883. *Über die Vererbung.* Jena: Fischer.

Dunn, L. C. 1965. *A Short History of Genetics.* New York: McGraw-Hill, pp. 44*ff.*

Strange, L. 1965. Plant cell differentiation. *Ann. Rev. Plant Physiol.* 16:119–40.

Stern, C. 1968. *Genetic Mosaics and Other Essays.* Cambridge: Harvard University Press, pp. 130*ff.*

CELL LINEAGE, THEORY OF

Virchow, R. L. C. 1860. *Cellular Biology as Based Upon Physiological and Pathological Histology.* Transl. by F. Chance. London: Churchill.

Maienschein, J. 1978. Cell lineage, ancestral reminiscence, and the biogenetic law. *J. Hist. Biol.* 11:129–58.

CELL RESPIRATION

Keilin, D. 1966. *The History of Cell Respiration and Cytochrome.* London: Cambridge University Press.

CELL THEORY

Schleiden, M. J. 1838. Beiträge zur Phytogenesis. *Johann Müllers Archiv für Anatomie und Physiologie,* 137–76.

Schwann, T. 1847. Microscopial researches into the accordance in the structure and growth of animals and plants. Transl. by H. Smith. London: Sydenham Society.

Virchow, R. 1855. Cellular pathology. (Transl.) *Disease, Life and Man: Selected Essays by Rudolph Virchow.* Edited by L. J. Rather. Stanford, Cal.: Stanford University Press, 1958.

Radl, E. 1930. *The History of Biological Theories.* Transl. by E. J. Hatfield. London: Oxford University Press, chap. 21.

Klein, M. 1936. *Histoire des Origines de la Théorie Cellulaire.* Paris.

Conklin, E. G. 1940. Cell and protoplasm concepts: historical account. *Cell and Protoplasm. Amer. Assoc. Adv. Sci. Publ.* 14:6–19.

Baker, J. R. 1949–55. The cell-theory: a restatement, history and critique. *Q. J. Microsc. Sci.* 89:103–25; 90:87–108; 93:157–90; 94:407–40; 96:449–81.

Gabriel, M. L., & S. Fogel. 1955. *Great Experiments in Biology.* Part I: The cell theory. Englewood Cliffs, N.J.: Prentice-Hall, pp. 1-22.

Causey, G. 1960. *The Cell of Schwann*. Edinburgh: Livingston; Baltimore: Williams & Wilkins.

Florkin, M. 1960. *Naissance et déviation de la théorie cellulaire dans l'oeuvre de Théodore Schwann*. Paris: Hermann.

Thompson, D'A. W. 1942. *On Growth and Form*. 2d ed. Cambridge: Cambridge University Press, pp. 341-45.

Davison, J. 1964. Animal organization as a problem in cell form. *Form and Strategy in Science*. Edited by J. R. Gregg & F. T. C. Harris. Dordrecht-Holland: Reidel, pp. 363-77.

Hershey, A. D. 1970. Genes and hereditary characteristics. *Nature* 226:697–700.

Kruta, V. 1971. J. E. Purkyně's contribution to the cell theory. *Clio. Med.* 6:109–20.

CELLS-FIRST HYPOTHESIS [origin of life]

Lehninger, A. L. 1970. *Biochemistry*. New York: Worth, p. 782.

CENTERS OF DIVERSITY *see* **AGE AND AREA**

CENTERS OF ORIGIN [biogeography]

Croizat, L., G. Nelson, & D. E. Rosen. 1974. Centers of origin and related concepts. *Syst. Zool.* 23:265–87.

McCoy, E. D., & K. L. Heck, Jr. 1976. Biogeography of corals, seagrasses, and mangroves: an alternative to the center of origin concept. *Syst. Zool.* 25:201–10.

Cf. GENE CENTERS

CENTRAL DOGMA [coding in protein synthesis]

Crick, F. H. C. 1958. On protein synthesis. *Symp. Soc. Exp. Biol.* 12:138–63.

Fox, S. W., K. Harada, & A. Vegotsky. 1959. Thermal polymerization of amino acids and a theory of biochemical origins. *Experientia* 15:81–84.

Watson, J. D. 1965. *Molecular Biology of the Gene*. New York: Benjamin, p. 297.

Jukes, T. H. 1966. *Molecules and Evolution*. New York: Columbia University Press, p. 187.

Olby, R. 1970. Francis Crick, DNA, and the central dogma. *Daedalus* 99:938–87.

——— . 1974. *The Path to the Double Helix*. London: Macmillan, pp. 432–34.

Judson, H. F. 1979. *The Eighth Day of Creation*. New York: Simon & Schuster, pp. 333*ff*.

CENTRAL-PATTERN-GENERATOR HYPOTHESIS [rhythmic behavior]

Wilson, D. M. 1964. The origin of the flight-motor command in grasshoppers. *Neural Theory and Modeling*. Edited by R. F. Reiss. Stanford, Cal.: Stanford University Press, pp. 331–45.

Hoyle, G. 1970. Cellular mechanisms underlying behavior-neuroethology. *Adv. Insect Physiol.* 7:349–444.

Brown, J. L. 1975. *The Evolution of Behavior*. New York: Norton, p. 504.

CENTRAL PLACE FORAGING [predation]

Orians, G. H., & N. E. Pearson. 1979. On the theory of central place foraging. *Analysis of Ecological Systems*. Edited by D. J. Horn, G. R. Stairs, & R. D. Mitchell. Columbus: Ohio State University Press, pp. 155–77.

CENTRIFUGAL SELECTION *see* **DISRUPTIVE SELECTION**

CENTRIFUGAL THEORY OF SPECIATION

= centrifugal speciation hypothesis

Brown, W. L., Jr. 1957. Centrifugal speciation. *Q. Rev. Biol.* 32:247–77.

CENTRIPETAL SELECTION *see* **NORMALIZING SELECTION**

CHAIN OF BEING [evolution; philosophy]

Robinet, J. B. R. 1768. *Considèrations philosophiques de la gradation naturelle des formes de l'être:* . . . Paris: Saillant.

Lovejoy, A. O. 1936. *The Great Chain of Being*. Cambridge: Harvard University Press, pp. 278–80.

Cf. SCALA NATURAE

CHAIN-REFLEX THEORY [of instinctive behavior]

Kalikov, T. J. 1975. History of Konrad Lorenz's ethological theory, 1927–1939. *Stud. Hist. Philos. Sci.* 6:331–41.

Cf. CYCLIC-REFLEX HYPOTHESIS

CHARACTER CONVERGENCE, THEORY OF [behavior; evolution]

Moynihan, M. H. 1968. Social mimicry: character convergence versus character displacement. *Evolution* 22:315–31.

Rosenzweig, M. L. 1968. Anecdotal evidence for the reality of character convergence. *Amer. Nat.* 102:491–92.

Cody, M. L. 1969. Convergent characteristics in sympatric species: a possible relation to interspecific competition and aggression. *Condor* 71:223–39.

———. 1973. Character convergence. *Annu. Rev. Ecol. Syst.* 4:189–211.

Cf. CONVERGENT EVOLUTION

CHARACTER DISPLACEMENT, THEORY OF [evolution]

Lack, D. 1947. *Darwin's Finches*. London: Cambridge University Press.

Brown, W. L., & E. O. Wilson. 1956. Character displacement. *Syst. Zool.* 5:49–64.

Grant, P. R. 1968. Bill size, body size, and the ecological adaptations of bird species to competitive situations on islands. *Syst. Zool.* 17:319–33.

———. 1972. Convergent and divergent character displacement. *Biol. J. Linn. Soc.* 4:39–68.

———. 1975. The classical case of character displacement. *Evol. Biol.* 8:237–337.

Kellogg, D. E. 1975. Character displacement in the radiolarian genus, *Eucrytidium*. *Evolution* 29:736–49.

Whalen, M. D. 1978. Reproductive character displacement and floral diversity in *Solanum* section *Androceras*. *Syst. Bot.* 3:77–86.

CHARGAFF RULES [DNA composition]

Chargaff, E. 1950. Chemical specificity of nucleic acids and mechanism of their enzymatic degradation. *Experientia* 6:201–09.

———. 1971. Preface to a grammar of biology. *Science* 172:637–42.

———. R. Lipshitz, C. Green, & M. E. Hodes. 1951. The composition of the deoxyribonucleic acid of salmon sperm. *J. Biol. Chem.* 192:223–30.

Watson, J. 1968. *The Double Helix*. New York: Atheneum, p. 125.

Olby, R. 1970. Francis Crick, DNA and the central dogma. *Daedalus* 99:938–87.

———. 1974. DNA before Watson-Crick. *Nature* 248:782–85.

Portugal, F. H., & J. S. Cohen. 1977. *A Century of DNA*. Cambridge: M.I.T. Press, p. 202.

Donohue, J. 1978. Review of *Heraclitean Fire: Sketches from a Life Before Nature*, by E. Chargaff. *Nature* 276:133–35.

Judson, H. F. 1979. *The Eighth Day of Creation*. New York: Simon & Schuster, pp. 96*ff*.

CHARLESWORTH-GIESEL HYPOTHESIS [selection]

Charlesworth, B., & J. T. Giesel. 1972. Selection in populations with overlapping generations. III. Relations between gene frequency and demographic variables. *Amer. Nat.* 106:388–402.

Giesel, J. T. 1972. Maintenance of genetic variability in natural populations—an alternative implication of the Charlesworth-Giesel hypothesis. *Amer. Nat.* 106:412–14.

CHEMICAL CORRELATION THEORY [plant development]

Sachs, J. von. 1880–82. Stoff und Form der Pflanzenorgane. *Arb. Bot. Inst. Würz.* 2:452–88, 689–718.

Wardlaw, C. W. 1952. *Phylogeny and Morphogenesis.* London: Macmillan, p. 439.

______ . 1965. *Organization and Evolution in Plants.* London: Longmans, pp. 26–27.

CHEMICAL HYPOTHESIS [of ion transport]

Slater, E. C. 1953. Mechanism of phosphorylation in the respiratory chain. *Nature* 172:975–78.

______ . 1966. Oxydative phosphorylation. *Comprehensive Biochemistry.* Edited by M. Florkin & E. H. Stotz. Amsterdam: Elsevier 14:327–96.

CHEMIOSMOTIC HYPOTHESIS [of ion transport]

Mitchell, P. 1966. Chemiosmotic coupling in oxidative and photosynthetic phosphorylation. *Biol. Rev.* 41:445–502.

______ . & J. Moyle. 1969. Estimation of membrane potential and pH difference across the cristae membrane of rat liver mitochondria. *Eur. J. Biochem.* 7:471–84.

Tien, H. T. 1974. *Bilayer Lipid Membranes (BLM).* New York: Dekker, pp. 381–84.

CHIASMATYPE THEORY [genetics]

= partial chiasmatype theory

Janssens, F. A. 1909. Spermatogénèse dans les batraciens. V. La théorie de la chiasmatypie. Nouvelle interprétation des cinèses de maturation. *Cellule* 25:389–411.

______ . 1924. Chiasmatypie dans les Insectes. *Cellule* 34:133–364.

Darlington, C. D. 1929. The significance of chromosome behaviour in polyploids for the theory of meiosis. John Innes Hort. Inst., London. *Conference on Polyploidy . . .* p. 42.

Kushev, V. V. 1974. *Mechanisms of Genetic Recombination.* New York: Consultants Bureau, chap. 1.

CHILD'S GRADIENT THEORY [of organization] *see* **GRADIENT THEORY**

CHILD'S THEORY [of plant development] *see* **GRADIENT THEORY**

CHIMERISM [genetics; development]

Ford, C. E. 1969. Mosaics and chimaeras. *Brit. Med. Bull.* 25:104–09.

Benirschke, K. 1970. Spontaneous chimerism in mammals: a critical review. *Curr. Top. Pathol.* 51:1–51.

CHOSSAT'S RULE [nutrition]

Chossat, C. 1843. Recherches experimentales sur l'inanition. *Mem. Acad. Sci.*, ser. 2, 8:438–640.

Kleiber, M. 1961. *The Fire of Life.* New York: Wiley, p. 37.

CHROMOSOMAL RACE [genetics]

Dobzhansky, T. 1944. Chromosomal races in *Drosophila pseudo-obscura* and *Drosophila persimilis. Carnegie Inst. Wash. Publ.* 554:47–144.

Powell, J. R., H. Levene, & T. Dobzhansky. 1973. Chromosomal polymorphism in *Drosophila pseudoobscura* used for diagnosis of geographic origin. *Evolution* 26:553–59.

CHROMOSOMAL SPECIATION

White, M. J. D. 1968. Models of speciation. *Science* 159:1065–70.

———. 1969. Chromosomal rearrangements and speciation in animals. *Annu. Rev. Genet.* 3:75–98.

Mayr, E. 1970. *Populations, Species, and Evolution.* Cambridge: Belknap Press of Harvard University Press, pp. 535 *ff.*

CHROMOSOME THEORY OF HEREDITY

Wilson, E. B. 1896. *The Cell in Development and Inheritance.* New York: Macmillan, chap. 9

Sutton, W. S. 1903. The chromosomes in heredity. *Biol. Bull.* 4:231–51.

Boveri, T. 1904. *Ergebnisse über die Konstitution der chromatischen Substanz des Zellkerns.* Jena: Fischer.

Morgan, T. H. 1914. The mechanism of heredity as indicated by the inheritance of linked characters. *Pop. Sci.* Jan., 1914, 1–16.

———. A. H. Sturtevant, H. J. Maller, & C. B. Bridges. 1915.

The Mechanism of Mendelian Heredity. New York: Holt, chap. 5.

Darlington, C. D. 1960. Chromosomes and the theory of heredity. *Nature* 187:892–95; *Smithson. Inst. Annu. Rep.*, 1961, 417–27.

Carlson, E. A. 1966. *The Gene: A Critical History*. Philadelphia: Saunders, pp. 30*ff*.

Baxter, E. B. 1976. Edmund B. Wilson as a preformationist: some reasons for his acceptance of the chromosome theory. *J. Hist. Biol.* 9:29–57.

______ . 1979. Mendel and meiosis. *J. Hist. Biol.* 12:137–73.

Voeller, B., ed. 1968. *The Chromosome Theory of Inheritance: Classic Papers in Development and Heredity*. New York: Appleton-Century-Crofts.

Roll-Hansen, N. 1978. *Drosophila* genetics: a reductionist research program. *J. Hist. Biol.* 11:159–210.

Cf. EPIGENESIS; FACTORIAL HYPOTHESIS; PARTICULATE CONCEPT

CHRONON CONCEPT [of circadian rhythms]

Ehret, C. F., & E. Trucco. 1967. Molecular models for the circadian clock. I. The chronon concept. *J. Theor. Biol.* 15:240–62.

Cf. BIOLOGICAL CLOCK

CICADA PRINCIPLE [ecology; evolution]

Beamer, R. H. 1931. Notes on the 17–year cicada in Kansas. *J. Kans. Entomol. Soc.* 4:53–58.

Lloyd, M., & H. S. Dybas. 1966. The periodical cicada problem. I and II. *Evolution* 20:133–49; 466–505.

Brown, J. 1975. *The Evolution of Behavior*. New York: Norton, pp. 143–44.

CIRCADIAN RHYTHM [evolution]

Pittendrigh, C. S. 1960. Circadian rhythms and the circadian organization of living systems. *Cold Spr. Harb. Symp. Quant. Biol.* 25:159–84.

Schweiger, E. 1964. Endogenous circadian rhythm and cytoplasm of *Acetabularia:* influence of the nucleus. *Science* 146:658–59.

Kramm, K.R. 1975. Entrainment of circadian activity rhythms in squirrels. *Amer. Nat.* 109:378-89.

Cf. BIOLOGICAL CLOCK

CIRCANNUAL RHYTHM [evolution]

Gwinner, E. 1971. A comparative study of circannual rhythms in warblers. *Biochronometry: Proceedings of a Symposium.*

Edited by M. Menaker. Washington, D.C.: National Academy of Sciences, pp. 405–27.
Stiles, F. G., & L. L. Wolf. 1974. A possible circannual molt rhythm in a tropical hummingbird. *Amer. Nat.* 108:341–54.

CIS-DOMINANT EFFECT [genetics]

Epstein, W., & J. R. Beckwith. 1968. Regulation of gene expression. *Annu. Rev. Biochem.* 37:411–36.

Cf. OPERON

CIS-TRANS EFFECT [genetics]

= LEWIS EFFECT

Lewis, E. B. 1945. The relation of repeats to position effect in *Drosophila melanogaster. Genetics* 30:137–66.
———. 1951. Pseudoallelism and gene evolution. *Cold Spr. Harb. Symp. Quant. Biol.* 16:159–74.
Benzer, S. 1957. The elementary units of heredity. *The Chemical Basis of Heredity.* Edited by W. D. McElroy & B. Glass. Baltimore: Johns Hopkins Press, pp. 70–93.

CLADISM *see* **PHYLOGENETIC SYSTEMATICS**

CLADOGENESIS [evolution]

Rensch, B. 1947. *Neuere Probleme der Abstammungslehre. Die transspezifische Evolution.* Stuttgart: Enke.
———. 1960. *Evolution Above the Species Level.* New York: Columbia University Press, chap. 6.
Simpson, G. G. 1949. Essay-review of recent works on evolutionary theory by Rensch, Zimmermann, and Schindewolf. *Evolution* 3:178–84.
Estabrook, G. F. 1972. Cladistic methodology: a discussion of the theoretical basis for the induction of evolutionary history. *Annu. Rev. Ecol. Syst.* 3:427–56.
Ashlock, P. D. 1974. The uses of cladistics. *Annu. Rev. Ecol. Syst.* 5:81–99.
Boucot, A. J. 1978. Community evolution and rates of cladogenesis. *Evol. Biol.* 11:545–655.

CLASSICAL SHOOT THEORY [of leaf origin]

Candolle, A. P. de. 1827. *Organographie Vegetale.* Tome 1. Paris: Deterville (transl. 1841 by H. Kingdon as: *Vegetable Organography.* London: Houlston & Stoneman).
Dickinson, T. A. 1978. Epiphylly in angiosperms. *Bot. Rev.* 44:181–232.

Cf. GOETHE–DE CANDOLLE THEORY

 CLASSICAL THEORY [of carpel evolution]

Hamshaw, T. H. 1934. The nature and origin of the stigma. *New Phytol.* 33:173–98.

Croizat, L. 1960. *Principia botanica.* 2 vols. Caracas: By the author, vol. 1, pp. 399*ff*, 569*ff*.

Cf. GOETHE–DE CANDOLLE THEORY

CLASSICAL THEORY [of the genetic structure of populations] = NEUTRAL THEORY [genetic fitness]

Muller, H. J. 1950. Our load of mutations. *Amer. J. Hum. Genet.* 2:111–76.

______ . & W. D. Kaplan. 1966. The dosage compensation of *Drosophila* and mammals as showing the accuracy of the normal types. *Genet. Res.* 8:41–59.

Dobzhansky, T. 1955. A review of some fundamental concepts and problems of population genetics. *Cold Spr. Harb. Symp. Quant. Biol.* 20:1–15.

Lewontin, R. C. 1974. *The Genetic Basis of Evolutionary Change.* New York: Columbia University Press, pp. 23, 194 *ff.*

Cf. BALANCE THEORY

CLEMENTSIAN ORGANISMIC VIEW [of the community] *see* **SUPRAORGANISMIC CONCEPT**

CLIMATIC MIGRATIONS, THEORY OF [plant geography]

Forbes, E. 1845. On the distribution of endemic plants, more especially those of the British Isles, considered with regard to geological changes. *Rept. Brit. Assoc. Adv. Sci.* 1845, 67–68.

______ . 1846. On the connexion between the distribution of the existing fauna and flora of the British Isles, and the geological changes which have affected their area, especially during the epoch of the northern drift. *Mem. Geol. Surv. U.K.* 1:336–432.

Good, R. 1974. *The Geography of the Flowering Plants.* 4th ed. London: Longman, p. 449.

CLIMATIC SELECTION [evolution]

Lamotte, M. 1951. Recherches sur la structure génétique des populations naturelles de *Cepaea memoralis* (L). *Bull. Biol. Fr. Belg.* (Suppl.) 35:1–238.

______ . 1959. Polymorphism of natural populations of *Cepaea memoralis* (L.) *Cold Spr. Harb. Symp. Quant. Biol.* 24:65–84.

Arnold, R. W. 1968. Climatic selection in *Cepaea memoralis* in the Pyrenees. *Phil. Trans. Roy. Soc. Lond. B.* 253:549–93.

———. 1969. The effects of selection by climate on the land snail *Cepaea memoralis*. *Evolution* 23:370–78.

Arnason, E., & P. R. Grant. 1976. Climatic selection in *Cepaea hortensis* at the northern limit of its range in Iceland. *Evolution* 30:499–508.

Cf. AREA EFFECTS

CLIMATIC STABILITY THEORY [of species diversity]

Klopfer, P. H. 1959. Environmental determinants of faunal diversity. *Amer. Nat.* 93:337–42.

Pianka, E. R. 1966. Latitudinal gradients in species diversity: a review of concepts. *Amer. Nat.* 100:33–46.

CLIMAX [ecology]

Clements, F. E. 1916. *Plant Succession. Carnegie Inst. Wash. Publ.* 242.

———. 1936. Nature and structure of the climax. *J. Ecol.* 24:252–84.

Tansley, A. G. 1935. The use and abuse of vegetational concepts and terms. *Ecology* 16:284–307.

Whittaker, R. H. 1953. A consideration of climax theory: the climax as a population and pattern. *Ecol. Monogr.* 23:41–78.

———. 1958. Recent evolution of ecological concepts in relation to the eastern forests of North America. *Fifty Years of Botany*. Edited by W. C. Steere. New York: McGraw-Hill, pp. 340–58.

Braun, E. L. 1958. The development of association and climax concepts. *Fifty Years of Botany*. Edited by W. C. Steere. New York: McGraw-Hill, pp. 329–39.

Selleck, G. W. 1960. The climax concept. *Bot. Rev.* 26:534–45.

Cf. CONTINUUM CONCEPT

CLIMBING THEORY [of sap ascent]

Malpighi, M. 1675–9. *Anatome Plantarum*. London: Joh. Martyn.

LeClerc du Sablon, M. 1910. Sur le mecanisme de la circulation de l'eau dans les plantes. *Rev. Gén. Bot.* 22:125–36.

Zimmerman, M. H. 1974. Long distance transport. *Plant Physiol.* 54:472–79.

CLINE [ecology; evolution]

= GEOGRAPHIC CHARACTER GRADIENTS

Gregor, J. W. 1939. Experimental taxonomy. IV. Population differentiation in North American and European sea plantains allied to *Plantago maritima* L. *New Phytol.* 38:293–322.

Huxley, J. S. 1942. *Evolution: The Modern Synthesis*. New York & London: Harper, pp. 206*ff*.

______. 1939. Clines: an auxillary method in taxonomy. *Bijdr. Dierk.* 27:491–520.

______. 1940. Towards the new systematics. *The New Systematics*. Edited by J. Huxley. Oxford: Clarendon Press, pp. 1–46.

Haldane, J. B. S. 1948. The theory of a cline. *J. Genet.* 48:277–84.

Endler, J. A. 1977. *Geographic Variation, Speciation, and Clines*. Princeton, N.J.: Princeton University Press.

CLONAL SELECTION THEORY [of antibody formation]

Burnet, F. M. 1957. A modification of Jerne's theory of antibody production using the concept of clonal selection. *Aust. J. Sci.* 20:67.

______. 1959. *The Clonal Selection Theory of Acquired Immunity*. London: Cambridge University Press.

Jerne, N. K. 1966. The natural selection theory of antibody formation: ten years later. *Phage and the Origins of Molecular Biology*. Edited by J. Cairns, G. S. Stent, & J. D. Watson. Cold Spring Harbor, N.Y.: Cold Spring Harbor Laboratory of Quantitative Biology, pp. 301–12.

Edelman, G. M. 1974. The problem of molecular recognition by a selective system. *Studies in the Philosophy of Biology*. Edited by F. J. Ayala & T. Dobzhansky. Berkeley: University of California Press, pp. 45–56.

Cf. SELECTION HYPOTHESES

CLOSED PROGRAM BEHAVIOR *see* **OPEN- AND CLOSED-PROGRAM BEHAVIOR**

CLUSTER CLONE THEORY [of mitochondrion origin]

Bogorad, L. 1975. Evolution of organelles and eukaryotic genomes. *Science* 188:891–98.

CLUTCH SIZE [avian evolution]

Lack, D. 1948. The significance of clutch size. *Ibis* 90:25–45.

Cody, M. L. 1966. A general theory of clutch size. *Evolution* 20:124–84.

Brockelman, W. Y. 1975. Competition, the fitness of offspring, and optimal clutch size. *Amer. Nat.* 109:677–99.

COACERVATE FORMATION THEORY [of multimolecular systems]

Oparin, A. I. 1957. *The Origin of Life on Earth*. New York: Academic Press, pp. 303*ff*.

Evreinova, T. N., T. W. Mamontova, V. N. Karnauhov, S. B. Stephanov, & U. R. Hrust. 1974. Coacervate systems and origin of life. *Orig. Life* 5:201–05.

COACTION THEORY [ecology]
= COACTION COMPASS

Haskell, E. F. 1949. A clarification of social science. *Main Currents in Modern Thought* 7:45–51.

Leary, R. A. 1976. Interaction geometry: an ecological perspective. *U.S. For. Serv. Gen. Tech. Rep. NC* no. 22.

Lidicker, W. Z. 1979. A clarification of interactions in ecological systems. *Bioscience* 29:475–77.

———. 1980. Coaction compasses (letter). *Bioscience* 30:76–77.

COADAPTATION

Dobzhansky, T. 1949. Observations and experiments on natural selection in *Drosophila*. *Proc. 8th Int. Cong. Genet.*, 1948 (*Hereditas,* suppl. vol.), pp. 210–24.

Mayr, E. 1963. *Animal species and evolution*. Cambridge: Belknap Press of Harvard University Press, pp. 271*ff*.

Cf. SYNORGANIZATION

COENOGENY [ecology; evolution]

Leppik, E. E. 1974. Phylogeny, hologeny, and coenogeny, basic concepts of environmental biology. *Acta Biotheor.* 23:170–93.

Cf. HOLOGENY

COEVOLUTION

Brues, C. T. 1920. The selection of food-plants by insects with special reference to lepidopterous larvae. *Amer. Nat.* 54:313–32.

Mode, C. J. 1958. A mathematical model for the co-evolution of obligate parasites and their hosts. *Evolution* 12:158–65.

Ehrlich, P. R., & P. H. Raven. 1964. Butterflies and plants: a study in co-evolution. *Evolution* 18:586–608.

Ashlock, P. D. 1974. The uses of cladistics. *Annu. Rev. Ecol. Syst.* 5:81–99.

Benson, W. W. 1975. Coevolution of plants and herbivores: passion flower butterflies. *Evolution* 29:659–80.

Gilbert, L., & P. Raven, eds. 1975. *Coevolution of Animals and Plants*. Austin: University of Texas Press.

Levin, S. A., & J. D. Udovic. 1977. A mathematical model of coevolving populations. *Amer. Nat.* 111:657–75.

Roughgarden, J. 1979. *Theory of Population Genetics and*

Evolutionary Ecology: An Introduction. New York: Macmillan, chap. 23.

Cf. GENE-FOR-GENE CONCEPT; MUTUALISM

COHESION THEORY [of sap ascent]

= Dixon-Askenazy cohesion theory

Askenazy, E. 1895. Ueber das Saftsteigen. *Verh. natur. med. Ver. Heidelb*, n.f. 5:325–45.

Dixon, H. H., & J. Joly. 1895. On the ascent of sap. *Phil. Trans. Roy. Soc. Lond. B*. 186:563–76.

______ . 1914. *Transpiration and the Ascent of Sap in Plants*. London: Macmillan.

Priestly, J. H. 1935. Sap ascent in the tree. *Sci. Prog.* 117:42–56.

Weevers, T. 1949. *Fifty Years of Plant Physiology*. Amsterdam: Scheltema & Holkema; Waltham, Mass.: Chronica Botanica, pp. 52–74.

Greenidge, K. N. H. 1957. Ascent of sap. *Annu. Rev. Plant Physiol.* 8:237–56.

Zimmermann, M. H. 1974. Long distance transport. *Plant Physiol.* 54:472–79.

COINCIDENCE THEORY [of evolutionary change]

Simpson, G. G. 1953. *The Major Features of Evolution*. New York: Columbia University Press, pp. 95*ff*.

Riedl, R. 1977. A systems-analytical approach to macro-evolutionary phenomena. *Q. Rev. Biol.* 52:351–70.

Cf. SALTATION; MUTATION THEORY

COLE'S RESULT [evolution]

Cole, L. C. 1954. The population consequences of life history phenomena. *Q. Rev. Biol.* 29:103–37.

Gadgil, M., & W. H. Bossert. 1970. Life historical consequences of natural selection. *Amer. Nat.* 104:1–24.

Bryant, E. 1971. Life-history consequences of natural selection: Cole's result. *Amer. Nat.* 105:75–76.

Charnov, E. L., & W. M. Schaffer. 1973. Life history consequences of natural selection: Cole's result revisited. *Amer. Nat.* 107:791–92.

Cf. LIFE HISTORY THEORY

COLLECTIVE HOMEOSTASIS *see* **GENETIC HOMEOSTASIS**

COMMUNAL BREEDING [bird behavior]
= cooperative breeding

Lack, D. 1968. *Ecological Adaptations for Breeding in Birds*. London: Methuen, p. 72.

Brown, J. L. 1974. Alternate routes to sociality in jays—with a theory for the evolution of altruism and communal breeding. *Amer. Zool.* 14:63–80.

COMMUNITY [ecology]

Clements, F. E. 1905. *Research Methods in Ecology*. Lincoln, Neb.: University Publishing.

———. 1928. *Plant Succession and Indicators*. New York: H. W. Wilson.

McIntosh, R. P. 1960. Natural order and communities. *Biologist* 42:55–62.

Whittaker, R. H. 1962. Classification of communities. *Bot. Rev.* 28:1–239.

———. 1975. *Communities and Ecosystems*. 2d ed. New York: Macmillan.

Klopfer, P. H. 1973. *Behavioral Aspects of Ecology*. 2d ed. Englewood Cliffs, N.J.: Prentice-Hall, chap. 5.

COMMUNITY MATRIX [ecology]

Levins, R. 1968. *Evolution in Changing Environments*. Princeton, N.J.: Princeton University Press, chap. 3.

Vandermeer, J. H. 1970. The community matrix and the number of species in a community. *Amer. Nat.* 104:73–83.

Neill, W. E. 1974. The community matrix and interdependence of the competition coefficients. *Amer. Nat.* 108:399–408.

COMMUNITY STRUCTURE *see* **BALANCE OF NATURE**

COMMUNITY-UNIT THEORY [ecology]

Whittaker, R.H. 1956. Vegetation of the Great Smoky Mountains. *Ecol. Monogr.* 26:1–80.

———. 1962. Classification of natural communities. *Bot. Rev.* 28:1–239.

———. 1967. Gradient analysis of vegetation. *Biol. Rev.* 42:207–64.

COMPARATIVE BIOLOGY, THEORY OF [philosophy; methodology]

Nelson, G. J. 1970. Outline of a theory of comparative biology. *Syst. Zool.* 19:373–84.

Urbani, V. C. B. 1972. Comments on Nelson's theory of comparative biology. *Z. Zool. Syst. Evolutionsforsch.* 10:152–54.

COMPARTMENT HYPOTHESIS [of insect development]

Lawrence, P. A., & G. Morata. 1976. The compartment hypothesis. *Symp. Roy. Entomol. Soc. Lond.* 8:132–49.

COMPENSATION *see* **BIOLOGICAL COMPENSATION; RUBNER'S LAW OF COMPENSATION**

COMPETITION [ecology; evolution]

Clements, F. E., & V. E. Shelford. 1939. *Bio-Ecology.* New York: Wiley, chap. 5.

de Wit, C. T. 1960. On competition. *Versl. Landbouwk. Onderz.* 66:8–82.

Klomp, H. 1961. The concepts "similar ecology" and "competition" in animal ecology. *Arch. Neerl. Zool.* 14:90–102.

Milne, A. 1961. Definition of competition among animals. *Mechanisms in Biological Competition.* Edited by F. L. Milthorpe. London: Cambridge University Press, pp. 40-61. (Also, *Symp. Soc. Exp. Biol.* 15:40–61.)

Egerton, F. N. 1968 (publ. 1971). The concept of competition in nature before Darwin. *Actes XIIe Congr. Int. Hist. Sci. 1968* 8:41–46.

McIntosh, R. P. 1970. Community, competition, and adaptation. *Q. Rev. Biol.* 45:259–80.

COMPETITION HYPOTHESIS [of species diversity]

Dobzhansky, T. 1950. Evolution in the tropics. *Amer. Sci.* 38:208–21.

Pianka, E. R. 1966. Latitudinal gradients in species diversity: a review of concepts. *Amer. Nat.* 100:33–46.

______ . 1974. *Evolutionary Ecology.* New York: Harper & Row, pp. 123*ff*, 241.

Ayala, F. J. 1970. Competition, coexistence, and evolution. *Essays in Evolution and Genetics in Honor of Theodosius Dobzhansky.* Edited by M. K. Hecht & W. C. Steere. New York: Appleton-Century-Crofts, pp. 121–58.

Menge, B. A., & J. P. Sutherland. 1976. Species diversity gradients: synthesis of the roles of predation, competition, and temporal heterogeneity. *Amer. Nat.* 110:351–69.

COMPETITIVE EXCLUSION [ecology]

= Gause's principle = Grinnell's axiom

Gause, G. F. 1934. *The Struggle for Existence.* Baltimore: Williams & Wilkins. (1964 reprint, New York: Hafner.)

Cole, L. C. 1960. Competitive exclusion. *Science* 132:348–49.

Hardin, G. 1960. The competitive exclusion principle. *Science* 131:1291–97.

Patten, B. C. 1961. Competitive exclusion. *Science* 134:1599–1601.

DeBach, P. 1966. The competitive displacement and coexistence principles. *Annu. Rev. Entomol.* 11:183–212.

Ayala, F. J. 1969. Experimental invalidation of the principle of competitive exclusion. *Nature* 224:1076–79.

Levin, S. A. 1970. Community equilibria and stability, and an extension of the competitive exclusion principle. *Amer. Nat.* 104:413–23.

Leak, W. B. 1972. Competitive exclusion in forest trees. *Nature* 236:461–63.

Forcier, L. K. 1975. Reproductive strategies and the co-occurrence of climax tree species. *Science* 189:808–09.

Kaplan, J. L., & J. A. Yorke. 1977. Competitive exclusion and nonequilibrium coexistence. *Amer. Nat.* 111:1030–36.

Diamond, J. M. 1978. Niche shifts and the rediscovery of interspecific competition. *Amer. Sci.* 66:322–31.

Krebs, C. 1978. *Ecology*. 2d ed. New York: Harper & Row, p. 225.

Huston, M. 1979. A general hypothesis of species diversity. *Amer. Nat.* 113:81–101.

Cf. LOTKA-VOLTERRA MODEL

COMPETITIVE SELECTION

Mather, K. 1969. Selection through competition *Heredity* 24:529–40.

Papentin, F. 1970. Kompetitive Selektion in einfachen Systemen. *Theor. Appl. Genet.* 40:218–25.

COMPETITIVE STRATEGY [in plants; evolution]

Grime, J. P. 1977. Evidence for the existence of three primary strategies in plants and its relevance to ecological and evolutionary theory. *Amer. Nat.* 111:1169–94.

Cf. COMPETITION

COMPILOSPECIES [systematics]

Harlan, J. R., & J. M. J. deWet. 1963. The compilospecies concept. *Evolution* 17:497–501.

COMPLEX ORGANISM [ecology]

Phillips, J. 1935. Succession, development, the climax, and the complex organism: an analysis of concepts. III. The complex organism: conclusions. *J. Ecol.* 23:488–508.

COMPLEXITY [evolution; ecology; philosophy]

Pringle, J. W. S. 1951. On the parallel between learning and evolution. *Behaviour* 3:174–215.

Bertalanffy, L. von. 1952. *Problems of Life*. New York: Wiley.

Bray, J. R. 1958. Notes toward an ecologic theory. *Evolution* 39:770–76.

Bronowski, J. 1974. New concepts in the evolution of complexity. *Bost. Stud. Philos. Sci.* 11:133–51.

COMPOSITE CHROMOSOME THEORY [of macronuclear organization]

Grell, K. G. 1953. Die Chromosomen von *Aulacantha scolymantha* Haeckel. *Arch. Protistenk.* 99:1–54.

———. & A. Ruthmann. 1964. Über die Karyologie des Radiolars *Aulacantha scolymantha* und die Feinstruktur seiner Chromosomen. *Chromosoma* 15:185–211.

Raikov, I. B. 1969. The macronucleus of ciliates. *Research in Protozoology*. Edited by T. T. Chen. Oxford: Pergamon, III:1–128.

Cf. SUBNUCLEAR HYPOTHESIS

COMPRESSION HYPOTHESIS [feeding behavior]

MacArthur, R. H., & E. R. Pianka. 1966. On optimal use of a patchy environment. *Amer. Nat.* 100:603–09.

———, & E. O. Wilson. 1967. *The theory of island biogeography*. Princeton, N.J.: Princeton University Press.

———. 1972. *Geographical Ecology*. New York: Harper & Row, p. 63.

Schoener, T. W. 1974. The compression hypothesis and temporal resource partitioning. *Proc. Natl. Acad. Sci.* 71:4169–72.

CONCENTRATION THEORY [of mammalian molar origin]

Bolk, L. 1913-19. *Odontologische Studien*. 3 vols. Jena: Fischer.

Peyer, B. 1968. *Comparative Odontology*. Chicago: University of Chicago Press, p. 182.

Dullemeijer, P. 1974. *Concepts and Approaches in Animal Morphology*. Assen, Netherlands: Van Gorcum, p. 16.

CONCOMITANTLY VARIABLE CODONS [genetics]

Fitch, W. M., & E. Markowitz. 1970. An improved method for determining codon variability in a gene and its application to the rate of fixation of mutations in evolution. *Biochem. Genet.* 4:579–93.

———, & E. Margoliash. 1970. The usefulness of amino acid

and nucleotide sequences in evolutionary studies. *Evol. Biol.* 4:67–109.
——— . 1971. Rate of change of concomitantly variable codons. *J. Mol. Evol.* 1:84–96.

CONCRESCENCE THEORY [of mammalian molar origin]

Kukenthal, W. 1889–93. *Vergleichend-anatomische und entwickelungsgeschichtliche Untersuchungen an Walthieren.* Jena: Fischer.

Neuville, H. 1932. Recherches comparatives sur la dentition des Cetodontes. *Ann. Sci. Nat. Zool.* 15:185–362.

Peyer, B. 1968. *Comparative Odontology.* Chicago: University of Chicago Press, p. 179.

Dullemeijer, P. 1974. *Concepts and Approaches in Animal Morphology.* Assen, Netherlands: Van Gorcum, p. 16.

CONDITIONAL HETEROSIS [genetics]

Röbbelen, G. 1957. Untersuchungen an strahleninduzierten Blattfarbmutanten von *Arabidopsis thaliana* (L.) Heynh. *Z. Indukt. Abstammungs-Vererbungsl.* 88:189–252.

Wills, C. 1973. In defense of naive pan-selectionism. *Amer. Nat.* 107:23–34.

——— . & L. Nichols. 1971. Single gene heterosis in *Drosophila* revealed by inbreeding. *Nature* 233:123–25.

——— . 1971. How genetic background masks single gene heterosis in *Drosophila. Proc. Natl. Acad. Sci.* 69:323–25.

Kidwell, J. F. 1974. Single-gene conditional heterosis? *Amer. Nat.* 108:883–86.

CONFLICT HYPOTHESIS [behavior]

Tinbergen, N. 1951. *The Study of Instinct.* Oxford: Clarendon Press, pp. 114*ff.*

Baerends, G. P. 1975. An evaluation of the conflict hypothesis as an explanatory principle for the evolution of displays. *Function and Evolution in Behavior.* Edited by G. P. Baerends, C. Beer, and A. Manning. Oxford: Clarendon Press, pp. 187–227.

CONSERVATION OF ORGANIZATION, PRINCIPLE OF [evolution]

Stebbins, G. L. 1969. *The Basis of Progressive Evolution.* Chapel Hill: University of North Carolina Press, pp. 124–26.

———. 1974. *Flowering Plants: Evolution above the Species Level.* Cambridge: Belknap Press of Harvard University Press, p. 24.

Cf. INTEGRATION OF THE GENOTYPE

CONSERVATIVE REGIONS, DOCTRINE OF [plant evolution]

= conservative characters or features

Bailey, I. W. 1912. The evolutionary history of the foliar ray in the wood of dicotyledons and its phytogenetic significance. *Ann. Bot.* 26:647–61.

Sporne, K. R. 1956. The phylogenetic classification of the angiosperms. *Biol. Rev.* 31:1–29.

Thorne, R. F. 1958. Some guiding principles of angiosperm phylogeny. *Brittonia* 10:72–77.

CONSTANT ANGLE, LAW OF [in shell formation]

Rhumbler, L. 1902. Die Doppelschalen von Orbitolites und anderer Foraminiferen. . . .*Arch. Protistenk.* 1:193–296.

Thompson, D'A. W. 1942. *On Growth and Form.* 2d ed. Cambridge: Cambridge University Press, pp. 862-63.

CONSTANT EXTINCTION, LAW OF *see* **RED QUEEN HYPOTHESIS**

CONSTRAINT RELATIONSHIPS [population genetics]

Gillois, M. 1966. La relation de dépendance entre gènes nonidentiques. *Ann. Biol. Anim. Biochim. Biophys.* 6:117–20.

______ . 1967. Les Lois conjointes des variables aléatoires génétiques. *Ann Génét.* 10:203–05.

Gallais, A. 1977. A general approach to dependance relationships among genes with some applications. *Proceedings of the International Conference on Quantitative Genetics, August 16-21, 1976.* Edited by E. Pollack, O. Kempthorne, & T. B. Bailey. Ames: Iowa State University Press, pp. 829–36.

CONTACT INHIBITION [cell biology]

Abercrombie, M., & E. J. Ambrose. 1962. The surface properties of cancer cells: a review. *Cancer Res.* 22:525–48.

CONTACT THEORY [of intercellular communication] *see* **NEURON THEORY**

CONTAGIOUS-DISEASE-BUFFER HYPOTHESIS [evolution of territoriality]

Carrick, R. 1963. Ecological significance of territory in the Australian magpie. *Gymnorhina tibicen. Proc. Int. Ornithol. Congr.* 13:740–53.

Verner, J. 1977. On the adaptive significance of territoriality. *Amer. Nat.* 111:769–75.

CONTAMINATION THEORY [of inheritance]

Castle, W. E. 1906. Yellow mice and gametic purity. *Science* 24:275–81.

Carlson, E. A. 1966. *The Gene: A Critical History*. Philadelphia: Saunders, pp. 23*ff*.

CONTEMPORANEOUS DISEQUILIBRIUM [ecology]

Richerson, P., R. Armstrong, & C. R. Goldman. 1970. Contemporaneous disequilibrium, a new hypothesis to explain the "paradox of the plankton." *Proc. Natl. Acad. Sci.* 67:1710–14.

Petersen, R. 1975. The paradox of the plankton: an equilibrium hypothesis. *Amer. Nat.* 109:35–49.

Cf. NONEQUILIBRIUM; PARADOX OF THE PLANKTON

CONTINENTAL DRIFT [biogeography]

Wegener, A. 1924. *The Origin of Continents and Oceans*. Transl. by J. G. A. Skerl. London: Methuen.

Blackett, P. M. S., E. Bullard, & S. K. Runcorn, eds. 1965. A symposium on continental drift. *Phil. Trans. Roy. Soc. Lond.* A 258:1–323.

Diety, R. S., & J. C. Holden. 1970. The breakup of Pangaea. *Sci. Amer.* 223(4):30–41.

Keast, A. 1971. Continental drift and the evolution of the biota on southern continents. *Q. Rev. Biol.* 46:335–78.

———. 1972. Continental drift and the evolution of the biota on southern continents. *Evolution, Mammals, and Southern Continents*. Edited by A. Keast, F. C. Erk, & B. Glass. Albany: State University of New York Press, pp. 23–87.

Schuster, R. M. 1972. Continental movements, "Wallace's line," and Indomalayan-Australasian dispersal of land plants: some eclectic concepts. *Bot. Rev.* 38:3–86.

Scientific American. 1972. *Continents Adrift*. San Francisco: Freeman.

Tarling, D. H., & S. K. Runcorn, eds. 1973. *The Implications of Continental Drift to the Earth Sciences*. New York: Academic Press.

Good, R. 1974. *The Geography of Flowering Plants*. 4th ed. London: Longmans, chap. 21.

Frankel, H. 1976. Alfred Wegener and the specialists. *Centaurus* 20:305–24.

———. 1979. The reception and acceptance of continental drift theory as a rational episode in the history of science. *The Reception Of Unconventional Science*. Edited by S. M. Mauskopf.Boulder, Colo.: Westview Press (AAAS Selected Symposium, #25), pp. 51–89.

Cf. PLATE TECTONICS

 CONTINUITY THEORY [of intercellular communication] *see* **RETICULAR THEORYCONTINUOUS MACROMOLECULE CONCEPT OF THE GENE**

Goldschmidt, R. B. 1938. *Physiological Genetics*. New York: McGraw-Hill, chap. 4.

Carlson, E. A. 1966. *The Gene: A Critical History*. Philadelphia: Saunders, pp. 128–30.

CONTINUOUS VARIATION *see* **ANCESTRAL INHERITANCE, LAW OF**

CONTINUUM CONCEPT [ecology]

Curtis, J. T., & R. P. McIntosh. 1951. An upland continuum in the prairie-forest border region of Wisconsin. *Ecology* 32:476–96.

Whittaker, R. H. 1951. A criticism of the plant association and climatic climax concepts. *Northwest Sci.* 25:17–31.

______ . 1957. Recent evolution of ecological concepts in relation to the eastern forests of North America. *Amer. J. Bot.* 44:197–206.

McIntosh, R. P. 1967. The continuum concept of vegetation. *Bot. Rev.* 33:130–87.

Dansereau, P., ed. 1968. The continuum concept of vegetation: responses. *Bot. Rev.* 34:253–332.

Cf. CLIMAX; GRADIENT ANALYSIS; INDIVIDUALISTIC CONCEPT

CONTINUUM CONCEPT [of water movement]

Gradmann, H. 1928. Untersuchungen über die Wasserverhältnisse des Bodens als Grundlage des Pflanzenwachstums. *Jahrb. Wiss. Bot.* 69:1–100.

van der Honert, T. H. 1948. Water transport as a catenary process. *Disc. Faraday Soc.* 3:146–53.

CONTRACTION-WAVE CONCEPT (OR FALLACY) [of vertebrate swimming]

Borelli, G. A. 1680. *De motu Animalium*. Rome: Bernaho.

Breder, C. A. 1926. The locomotion of fishes. *Zoologica* (N.Y.) 4:159–297.

Blight, A. R. 1977. The muscular control of vertebrate swimming movements. *Biol. Rev.* 52:181–218.

CONTROL THEORY [of animal behavior]

Bayliss, L. E. 1966. *Living Control Systems*. London: English Universities Press.

Milhorn, H. T., Jr. 1966. *The Application of Control Theory to Physiological Systems.* Philadelphia: Saunders.

Milsum, J. H. 1966. *Biological Control Systems Analysis.* New York: McGraw-Hill.

McFarland, D. J. 1971. *Feedback Mechanisms in Animal Behaviour.* London: Academic Press.

CONVERGENCE *see* **CONVERGENT EVOLUTION**

CONVERGENT EVOLUTION

Mayr, E. 1963. *Animal Species and Evolution.* Cambridge: Belknap Press of Harvard University Press, p. 609.

Mooney, H. A., & E. L. Dunn. 1970. Convergent evolution of Mediterranean-climate evergreen Sclerophyll shrubs. *Evolution* 24:292–303.

Böcher, T. W. 1977. Convergence as an evolutionary process. *Bot. J. Linn. Soc.* 75:1–19.

Mooney, H. A., ed. 1977. *Convergent Evolution in Chile and California.* Stroudsburg, Pa.: Dowden, Hutchinson & Ross.

Cody, M. L., & H. A. Mooney. 1978. Convergence versus nonconvergence in Mediterranean-climate ecosystems. *Annu. Rev. Ecol. Syst.* 9:265–321.

Peet, R. K. 1978. Ecosystem convergence. *Amer. Nat.* 112:441–44.

Cf. PARALLEL EVOLUTION; MIMICRY

COOPERATION [ecology; behavior]

Kropotkin, P. 1915. *Mutual Aid as a Factor in Evolution.* London: Heinemann.

Clements, F. E., & V. E. Shelford. 1939. *Bio-ecology.* New York: Wiley, chap. 5.

Allee, W. C. 1951. *Cooperation Among Animals with Human Implications.* New York: Henry Schuman.

COOPERATIVE BREEDING *see* **COMMUNAL BREEDING**

COOPERATIVE GROUP [ecology; evolution; behavior]

Walker, I., & R. M. Williams. 1976. The evolution of the cooperative group. *Acta Biotheor.* 25:1–43.

COPE-OSBORN THEORY OF TRITUBERCULY [mammalian molar origin]

Cope, E. D. 1874. On the homologies and origin of the types of molar teeth of mammalia *Educabilia. J. Acad. Nat. Sci. Phila.* (2) 8:71–89.

Osborn, H. F. 1888. The evolution of mammalian molars to and from the triangular type. *Amer. Nat.* 22:1067–79.

Gregory, W. K. 1934. A half century of trituberculy; the Cope-Osborn theory of dental evolution. *Proc. Amer. Philos. Soc.* 73:169–317.

Butler, P. M. 1941. A theory of the evolution of mammalian molar teeth. *Amer. J. Sci.* 239:421–50.

Peyer, B. 1968. *Comparative Odontology*. Chicago: University of Chicago Press, pp. 185*ff*.

Hopson, J. A., & A. W. Crompton. 1969. Origin of Mammals. *Evol. Biol.* 3:15–72.

Dullemeijer, P. 1974. *Concepts and Approaches in Animal Morphology*. Assen, Netherlands: Van Gorcum, p. 16.

COPE'S LAW (OR RULE) OF THE UNSPECIALIZED [evolution]

Cope, E. D. 1885. On the evolution of the vertebrata. *Amer. Nat.* 19:140–48, 234–47, 341–53.

______ . 1896. *The Primary Factors of Organic Evolution*. Chicago: Open Court, p. 172.

Amadon, D. 1943. Specialization and evolution. *Amer. Nat.* 77:133–41.

Rensch, B. 1960. *Evolution Above the Species Level*. New York: Columbia University Press, pp. 206–12.

Mayr, E. 1963. *Animal Species and Evolution*. Cambridge: Belknap Press of Harvard University Press, p. 595.

Bonner, J. T. 1965. *Size and Cycle*. Princeton, N.J.: Princeton University Press, pp. 178–81.

Stanley, S. M. 1973. An explanation for Cope's rule. *Evolution* 27:1–26.

COPY-CHOICE HYPOTHESIS [genetic recombination]

Lederberg, J. 1955. Recombination mechanisms in bacteria. *J. Cell. Comp. Physiol.* 45 (Suppl. 2):75–107.

Delbruck, M., & G. S. Stent. 1957. On the mechanism of DNA replication. *The Chemical Basis of Heredity*. Edited by W. D. McElroy & B. Glass. Baltimore: Johns Hopkins Press, pp. 699–736.

Kushev, V. V. 1974. *Mechanisms of Genetic Recombination*. New York: Consultants Bureau, chap. 1.

CORRELATED RESPONSES THEORY [selection; genetics]

Falconer, D. S. 1954. Validity of the theory of genetic correlation. An experimental test with mice. *J. Hered.* 45:42–44.

______ . 1960. *Introduction to Quantitative Genetics*. Edinburgh: Oliver & Boyd, chap. 19.

Clayton, G. A., G. R. Knight, J. A. Morris, & A. Robertson. 1957. An experimental check on quantitative genetical theory. III. Correlated responses. *J. Genet.* 55:171–80.

CORRELATION, DOCTRINE OF [plant evolution]

Frost, F. H. 1930. Serialization in secondary xylem of dicotyledons. I. Origin of vessels. *Bot. Gaz.* 89:67–94.

Sporne, K. R. 1956. The phylogenetic classification of the angiosperms. *Biol. Rev.* 31:1–29.

CORRESPONDENCE, LAW OF
= PARALLELISM, LAW OF

Urmantsev, Y. A. 1968. Polymorphism and isomorphism in living and inanimate nature. (Russ.) *Vopr. Filos.* 12:77–88.

Meyen, S. V. 1973. Plant morphology in its nomothetical aspects. *Bot. Rev.* 39:205–60.

COST OF NATURAL SELECTION *see* **HALDANE'S DILEMMA**

COUNTERADAPTATION HYPOTHESIS [of differences in egg production]

Ricklefs, R. E. 1970. Clutch size in birds: outcome of opposing predator and prey adaptations. *Science* 168:599–600.

Price, P. W. 1974. Strategies for egg production. *Evolution* 28:76–84.

COUPLING-REPULSION EFFECT [restricting overdominance; genetics]

Mukai, T., & T. Yamazaki. 1964. Position effect of spontaneous mutant polygenes controlling viability in *Drosophila melanogaster*. *Proc. Jap. Acad.* 40:840–45.

———. 1968. The genetic structure of natural populations of *Drosophila melanogaster*. V. Coupling-repulsion effect of spontaneous mutant polygenes controlling viability. Genetics 59:513–35.

COVARIONS *see* **CONCOMITANTLY VARIABLE CONDONS**

CRABTREE EFFECT [inhibition of respiration by glycolysis]

Crabtree, H. G. 1929. Observations on the carbohydrate metabolism of tumors. *Biochem. J.* 23:536–45.

Racker, E. 1956. Carbohydrate metabolism in ascites tumor cells. *Ann. N.Y. Acad. Sci.* 63:1017–21.

______ . 1965. *Mechanisms in Bioenergetics.* New York: Academic Press, p. 205.

Cf. PASTEUR EFFECT

CREATIONISM *see* **SPECIAL CREATION**

CROSSOVER FIXATION THEORY [homogeneity of rRNA repeats]

Smith, G. P. 1973. Unequal crossover and the evolution of multigene families. *Cold Spr. Harb. Symp. Quant. Biol.* 38:507–13.

CROWDING EFFECT [in tapeworms]

Read, C. P. 1951. The "crowding effect" in tapeworm infections. *J. Parasit.* 37:174–78.

______ . 1959. The role of carbohydrates in the biology of cestodes. VIII. Some conclusions and hypotheses. *Exp. Parasit.* 8:365–82.

Roberts, L. S. 1966. Developmental physiology of cestodes. I. Host dietary carbohydrate and the "crowding effect" in *Hymenolepis diminuta. Exp. Parasit.* 18:305–10.

CRYPTIC SPECIES *see* **SIBLING SPECIES**

CRYPTIC STRUCTURAL HYBRIDITY HYPOTHESIS [chromosome repatterning]

Stebbins, G. L. 1950. *Variation and Evolution in Plants.* New York: Columbia University Press, pp. 222*ff.*

______ . 1971. *Chromosomal Evolution in Higher Plants.* London: Arnold, pp. 117*ff.*

CRYPTOBIOSIS *see* **ANABIOSIS**

CRYSTALLIZATION HYPOTHESIS [of cell growth]

Schwann, T. 1847. Microscopial researches into the accordance of the structure and growth of animals and plants. Transl. by H. Smith. London: Sydenham Society, pp. 220–57.

Galaty, D. H. 1974. The philosophical basis of mid-nineteenth century German reductionism. *J. Hist. Med.* 29:295–316.

CURSORIAL PREDATOR THEORY [of bird flight origin]

Ostrom, J. H. 1974. *Archaeopteryx* and the origin of flight. *Q. Rev. Biol.* 49:27–47.

______ . 1979. Bird flight: how did it begin? *Amer. Sci.* 67:46–56.

CURSORIAL THEORY [of bird flight origin]

Nopcsa, F. 1907. Ideas on the origin of flight. *Proc. Zool. Soc. Lond.* 1907:223–36.

Ostrom, J. H. 1974. *Archaeopteryx* and the origin of flight. *Q. Rev. Biol.* 49:27–47.

CYCLIC-REFLEX HYPOTHESIS [of rhythmic behavior patterns]

Wilson, D. M. 1966. Central nervous mechanisms for the generation of rhythmic behavior in arthropods. *Symp. Soc. Exp. Biol.* 20:199–228.

Hoyle, G. 1970. Cellular mechanisms underlying behavior-neuroethology. *Adv. Insect Physiol.* 7:349–444.

Brown, J. L. 1975. *The Evolution of Behavior.* New York: Norton, p. 504.

CYCLIC SUCCESSION [ecology]

Watt, A.S. 1947. Pattern and process in the plant community. *J. Ecol.* 35:1–22.

Connell, J. H., & R. O. Slayter. 1977. Mechanisms of succession in natural communities and their role in community stability and organization. *Amer. Nat.* 11:119–44.

Cf. SUCCESSION

CYCLICAL SELECTION *see* **ENDOCYCLIC SELECTION**

CYCLOL THEORY [of protein structure]

Wrinch, D. M. 1938. Is there a protein fabric? *Cold Spr. Harb. Symp. Quant. Biol.* 6:112–39.

———. 1965. *Chemical Aspects of Polypeptide Structure and the Cyclol Theory.* New York: Plenum.

CYCLOMORPHOSIS

Coker, R. E. 1939. The problem of cyclomorphosis in Daphnia. *Q. Rev. Biol.* 14:137–48.

Brooks, J. L. 1946. Cyclomorphosis in Daphnia. I. An analysis of *D. retrocurva* and *D. galeata. Ecol. Monogr.* 16:409–47.

Hutchinson, G. E. 1967. *A Treatise on Limnology.* Vol. II. New York: Wiley, chap. 26.

Dodson, S. I. 1974. Adaptive change in plankton morphology in response to size-selective predation: a new hypothesis of cyclomorphosis. *Limnol. Oceanogr.* 19:721–29.

CYTOCATALYTIC EVOLUTION

Lewis, W. H. 1967. Cytocatalytic evolution in plants. *Bot. Rev.* 33:105–15.

DALCQ-PASTEELS THEORY [of early embryonic fields]

Dalcq, A. M. 1938. *Form and Causality in Early Development.* Cambridge: Cambridge University Press, pp. 77–102.

______ . & J. Pasteels. 1937. Une conception nouvelle des bases physiologiques de la morphogénèse. *Arch. Biol.* 48:669–710.

Needham. J. 1942. *Biochemistry and Morphogenesis.* Cambridge: Cambridge University Press, p. 220.

DANIELLI MODEL [of membrane structure]

Danielli-Davson hypothesis

Danielli, J. F., & H. Davson. 1935. A contribution to the theory of permeability of thin films. *J. Cell. Comp. Physiol.* 5:495–508.

Stoeckenius, W. 1970. Electron microscopy of mitochondrial and model membranes. *Membranes of Mitochondria and Chloroplasts.* Edited by E. Racker. New York: Van Nostrand Reinhold, pp. 53–90.

Cf. UNIT-MEMBRANE THEORY

DARLING EFFECT [ecology]

Darling, F. F. 1938. *Bird Flocks and the Breeding Cycle.* New York: Cambridge University Press.

Fisher, J. 1954. Evolution and bird sociality. *Evolution as a Process.* Edited by J. Huxley, A. C. Hardy, & E. B. Ford. London: Allen & Unwin, pp. 71–83.

DARLINGTON RULE [of fertility of allopolyploids]

Darlington, C. D. 1928, Studies in *Prunus*, I and II. *J. Genet.* 19:213–56.

______ . 1932. The control of the chromosomes by the genotype and its bearing on some evolutionary problems. *Amer. Nat.* 66:25–51.

Dobzhansky, T. 1937. *Genetics and the Origin of Species.* New York: Columbia University Press, pp. 207, 290–93.

DARWINIAN SEXUAL SELECTION *see* **SEXUAL SELECTION**

DARWIN'S ABOMINABLE MYSTERY [angiosperm origin]

Darwin, C. 1879. Letter to J. D. Hooker, July 22. *More Letters of Charles Darwin.* Edited by F. Darwin. New York: Appleton, 1903, pp. 20–21.

Radford, A. E., W. C. Dickison, J. R. Massey, & C. Ritchie Bell. 1974. *Vascular Plant Systematics*. New York: Harper & Row, pp. 553–60.

Doyle, J. A. 1978. Origin of angiosperms. *Annu. Rev. Ecol. Syst.* 9:365–92.

Cf. NEOTONY

DARWIN'S THEORY OF TRANSMUTATION [of species]

de Beer, G. 1960. Darwin's notebooks on transmutation of species. *Bull. Brit. Mus. Hist. Ser.* 2:27–73.

——— , & M. J. Rowlands. 1961. Darwin's notebook on transmutation of species. Addenda and corrigenda. *Bull. Brit. Mus. Hist. Ser.* 2:187–200.

Grinnell, G. 1974. The rise and fall of Darwin's first theory of transmutation. *J. Hist. Biol.* 7:259–73.

Herbert, S. 1974. The place of man in the development of Darwin's theory of transmutation. Part I. To July 1837. *J. Hist. Biol.* 7:217–58.

DAUER-MODIFICATION [genetics]

Jollos, V. 1921. Experimentelle protostenstudien. I. *Arch. Protistenk.* 43:1–222.

——— . 1924. Untersuchungen über Vererbung und Variabilität bei Arcellen. *Arch. Protistenk.* 49:307–74.

——— . 1939. Grundbegriffe der Vererbungslehre, insbesondere Mutation, Dauermodifikation, Modifikation. *Handbuch der Vererbungswissenschaft*. Edited by E. Bauer & M. Hartmann. Berlin: gibrüder Borntraeger, Bd. 4.

Goldschmidt, R. B. 1955. *Theoretical Genetics*. Berkeley: University of California Press, pp. 201–05.

DEAR ENEMY PHENOMENON [behavior]

Fisher, J. 1954. Evolution and bird sociality. *Evolution as a Process*. Edited by J. Huxley. A. C. Hardy, & E. B. Ford. London: Allen & Unwin, pp. 71–83.

Wilson, E. O. 1975. *Sociobiology*. Cambridge: Belknap Press of Harvard University Press, pp. 273–74.

DECEIT [animal behavior]

= deception

Wallace, B. 1973. Misinformation, fitness, and selection. *Amer. Nat.* 107:1–7.

Otte, D. 1975. On the role of intraspecific deception. *Amer. Nat.* 109:239–42.

Cf. MIMICRY

DEFECTIVE COPYING *see* **GENE CONVERSION**

DEFICIENCY DISEASE CONCEPT [nutrition]

Follis, R. H., Jr. 1958. *Deficiency Disease*. Springfield, Ill.: Thomas.

______. 1960. Cellular pathology and the development of the deficiency disease concept. *Bull. Hist. Med.* 34:291–317.

DEGREE DAY CONCEPT [ecology]

= degree day plant phenology interaction

Wang, J. Y. 1960. A critique of the heat unit approach to plant response studies. *Ecology* 41:785–89.

Idso, S., & R. J. Reginato. 1978. Extending the "degree day" concept of plant phenological development to include water stress effects. *Ecology* 59:431–33.

DELAYED RETURN BENEFIT [animal behavior]

Trivers, R. L. 1971. The evolution of reciprocal altruism. *Q. Rev. Biol.* 46:35–57.

McKaye, K. R. 1977. Defense of a predator's young by a herbivorous fish: an unusual strategy. *Amer. Nat.* 111:301–15.

DEMAND THEORY [of adaptive growth] *see* **FUNCTIONAL DEMAND THEORY**

DEME [population genetics]

Gilmour, J. S. L., & J. W. Gregor. 1939. Demes. A suggested new terminology. *Nature* 144:333–34.

Lewontin, R. C., & L. C. Dunn. 1960. The evolutionary dynamics of a polymorphism in the house mouse. *Genetics* 45:705–22.

Reimer, J. D., & M. L. Petras. 1967. Breeding structure of the house mouse, *Mus musculus*, in a population cage. *J. Mammal.* 48:88–99.

Cf. INTERDEME SELECTION

DEMOCRATIC GENE CONVERSION THEORY [evolution]

Edelman, G. M. 1970. The structure and genetics of antibodies. *The Neurosciences: Second Study Program.* Edited by F. O. Schmitt. New York: Rockefeller University Press, pp. 885–900.

______, & J. A. Gally, 1970. Arrangement and evolution of eukaryotic genes. *Ibid.*, pp. 962–72.

DEMOGRAPHY [ecology]

Brass, W., ed. 1971. *Biological Aspects of Demography.* New York: Barnes & Noble.

Williamson, M. H. 1972. *The Analysis of Biological Populations*. London: Arnold.

Harper, J. L., & J. White. 1974. The demography of plants. *Annu. Rev. Ecol. Syst.* 5:419–63.

DENSITY DEPENDENCE [population control]

Smith, H. S. 1935. The role of biotic factors in the determination of population densities. *J. Econ. Entomol.* 28:873–98.

Milne, A. 1957. Theories of natural control of insect populations. *Cold Spr. Harb. Symp. Quant. Bio.* 22:253–71.

Cf. PERFECT DENSITY DEPENDENCE

DENSITY DEPENDENT MORTALITY *see* **ECOLOGICAL DEATH**

DENSITY-DEPENDENT SELECTION [evolution]

Roughgarden, J. 1971. Density-dependent natural selection. *Ecology* 52:453–68.

Clarke, B. 1972. Density-dependent selection. *Amer. Nat.* 106:1–13.

Bulmer, M. G. 1974. Density-dependent selection and character displacement. *Amer. Nat.* 108:45–58.

Cf. ALLEE EFFECT; FREQUENCY-DEPENDENT SELECTION

DENSITY-LIMITATION HYPOTHESIS [evolution of territoriality]

Howard, H. E. 1920. *Territory in Bird Life*. London: Murray.

Brown, J. L. 1969. Territorial behavior and population regulation in birds. *Wilson Bull.* 81:293–329.

Verner, J. 1979. On the adaptive significance of territoriality. *Amer. Nat.* 111:769–75.

DEPLETED SPECIES [biogeography]

Turesson, G. 1927. Contributions to the gen-ecology of glacial regions. *Hereditas* 9:81–101.

Stebbins, G. L. 1942. The genetic approach to problems of rare and endemic species. *Madroño* 6:241–58.

DEPRESSOR EFFECT [of bacteriophage reproduction]

Delbrück, M. 1945. Interference between bacterial viruses. III. The mutual exclusion effect and the depressor effect. *J. Bacteriol.* 50:151–70.

DETERMINATION, CONCEPT OF [morphogenesis]

Roux, W. 1905. Die Entwicklungsmechanik, ein neuer Zweig der biologischen Wissenschaft. *Vorträge und Aufsätze über*

Entwicklungsmechanik der Organismen. Edited by W. Roux. Leipzig: Engelmann, vol. I.

Waddington, C. H. 1956. *Principles of Embryology*. London: Allen & Unwin, pp. 411*ff*.

Hadorn, E. 1965. Problems of determination and transdetermination. *Brookh. Symp. Biol.* 18:148–61.

Cf. PERFORMATION

DEVELOPMENT [ecology]

Tansley, A. G. 1920. The classification of vegetation and the concept of development. *J. Ecol.* 8:118–44.

Phillips, J. 1935. Succession, development, the climax, and the complex organism: an analysis of concepts. Part II. Development and the climax. *J. Ecol.* 23:210–46.

DEVELOPMENT, LAW OF *see* **VON BAER'S LAWS**

DEVELOPMENTAL CANALIZATION *see* **CANALIZATION**

DEVELOPMENTAL FLEXIBILITY *see* **DEVELOPMENTAL HOMEOSTASIS**

DEVELOPMENTAL HOMEOSTASIS [genetics; evolution]
= developmental flexibility = phenotypic flexibility

Thoday, J. M. 1953. Components of fitness. *Symp. Soc. Exp. Biol.* 7:96–113.

Lerner, I. M. 1954. *Genetic Homeostasis*. New York: Wiley.

Dobzhansky, T. 1970. *Genetics of the Evolutionary Process*. New York: Columbia University Press, chap. 2.

Bruckner, D. 1976. The influence of genetic variability on wing symmetry in honeybees *(Apis mellifera)*. *Evolution* 30:100–08.

DEVIATION RULE [evolution]

Hennig, W. 1966. *Phylogenetic Systematics*. Urbana: University of Illinois Press, pp. 59, 207.

Løvtrup, S. 1974. *Epigenetics*. New York: Wiley, p. 469.

DIALECTIC THEORY OF BIOGENESIS

Körner, V. 1975. Zur dialektischen Theorie der Biogenese. *Biol. Zblt.* 94:249–76.

DIALLEL CROSSES [genetics]

Schmidt, J. 1919. La valeur de l' individu à titre de générateur appréciée suivant la méthode du croisement diallèle. *C. R. Trav. Lab. Carlsb.* 14, no. 633.

Hayman, B. I. 1954, 1958. The theory and analysis of diallel crosses. I and II. *Genetics* 39:789–809; 43:63–85.

Hinklemann, K. 1977. Diallel and multi-cross designs: what do they achieve? *Proceedings of the International Conference on Quantitative Genetics,* Aug. 16-26, 1976. Edited by E. Pollack, O. Kempthorne, & T. B. Bailey. Ames: Iowa State University Press, pp. 659–76.

DIFFERENTIAL ADHESION HYPOTHESIS [cell biology]

Steinberg, M. S. 1963. Reconstruction of tissues by dissociated cells. *Science* 141:401–08.

———. 1964. The problem of adhesive selectivity in cellular interactions. *Cellular Membranes in Development.* Edited by M. Locke. New York: Academic Press, pp. 321–66.

Jacobson, M. 1970. Development, specification, and diversification of neuronal connections. *The Neurosciences: Second Study Program.* Edited by F. O. Schmitt. New York: Rockefeller University, pp. 116–29.

Johnson, K. E. 1974. Gastrulation and cell interactions. *Concepts of Development.* Edited by J. Lash and J. R. Whittaker. Stamford, Conn.: Sinauer Associates, pp. 128–48.

DIFFERENTIAL NICHE EXPLOITATION [ecology, evolution]

Levin, D. A. 1974. Spatial segregation of pins and thrums in populations of *Hedyotis nigricans. Evolution* 28:648–55.

DIFFERENTIATION [physiology; morphogenesis]

Bonner, J. T. 1952. *Morphogenesis: An Essay on Development.* Princeton, N.J.: Princeton University Press, chap. 7.

Beerman, W. 1964. *Control or differentiation at the chromosomal level. Differentiation and Development.* A Symposium of the New York Heart Association. Boston: Little, Brown, pp. 49–61.

Heslop-Harrison, J. 1967. Differentiation. *Annu. Rev. Plant Physiol.* 18:325–48.

DIFFERENTIATION, THEORY OF [evolution; biogeography]
= Guppy's theory of differentiation

Guppy, H. B. 1903–06. *Observations of a Naturalist in the Pacific.* 2 vols. London & New York: Macmillan.

———. 1919. Plant distribution from the standpoint of an idealist. *J. Linn. Soc. Lond. Bot.* 44:439–72.

Willis, J. C. 1940. *The Course of Evolution, by Differentiation or Divergent Mutation Rather Than by Selection.* Cambridge: Cambridge University Press; New York: Macmillan.

Kurten, B. 1958. A differentiation index, and a new measure of evolutionary rates. *Evolution* 12:146–57.

Ehrlich, P. R., & P. H. Raven. 1969. Differentiation of populations. *Science* 165:1228–32.

Good, R. 1974. *The Geography of the Flowering Plants*. 4th ed. London: Longman, pp. 38–40.

DIFFERENTIATION THEORY [mammalian molar origin]

Peyer, B. 1968. *Comparative Odontology*. Chicago: University of Chicago Press, p. 185.

Cf. COPE-OSBORN THEORY

DIMERE THEORY *see* **CONCENTRATION THEORY**

DIRECT INTERACTION THEORY [origin of the genetic code]

Woese, C. 1965. Order in the genetic code. *Proc. Natl. Acad. Sci.* 54:71–75.

______ . 1969. Models for the evolution of codon assignments. *J. Mol. Biol.* 43:235–40.

Fitch, W. M. 1973. Aspects of molecular evolution. *Annu. Rev. Genet.* 7:343–80.

DIRECTED CORRECTION HYPOTHESIS [genetic recombination]

Kushev, V. V. 1974. *Mechanisms of Genetic Recombination*. New York: Consultants Bureau, chap. 5.

______ . 1975. Interpretation of intragenic recombination phenomena in terms of the principles of directional correction. *Molecular Mechanisms of Genetic Processes*. Edited by N. P. Dubinin & D. M. Gol'dfarb. Transl. by A. Mercado. New York: Wiley, pp. 285–89.

DIRECTIONAL SELECTION [evolution]

Schmalhausen, I. I. 1949. *Factors of Evolution: The Theory of Stabilizing Selection*. Philadelphia: Blakiston.

Mather, K. 1953. The genetical structure of populations. *Symp. Soc. Exp. Biol.* 7:66–95.

Camin, J. H., & P. R. Ehrlich. 1958. Natural selection in water snakes (*Natrix sipedon* L.) on islands in Lake Erie. *Evolution* 12:504–11.

Fox, S. F. 1975. Natural selection in morphological phenotypes of the lizard *Uta stansburiana*. *Evolution* 29:95–107.

Frazetta, T. H. 1975. *Complex Adaptations in Evolving Populations*. Sunderland, Mass.: Sinauer Associates, pp. 34-37.

DISASSORTATIVE MATING [genetics]

Sheppard, P. M. 1952. A note on non-random mating in the moth *Panaxia dominula* (L.). *Heredity* 6:239–41.

Lowther, J. K. 1961. Polymorphism in the white throated sparrow, *Zonotrichia albicollis* (Gmelin). *Can J. Zool.* 39:281–92.

Naylor, A. F. 1962. Mating systems which could increase heterozygosity for a pair of alleles. *Amer. Nat.* 96:51–60.

Smith, D. A. S. 1973. Negative non-random mating in the polymorphic butterfly *Danaus chrysippus* in Tanzania. *Nature* 242:131–32.

Borowsky, R., & K. D. Kallman. 1976. Patterns of mating in natural populations of *Xiphophorus* (Pisces: Poeciliidae). I. *X. maculatus* from Belize and Mexico. *Evolution* 30:693–706.

DISCHARGE HYPOTHESIS *see* **ELECTRICAL TRANSMISSION THEORY**

DISCONTINUOUS VARIATION [genetics]

Bateson, W. 1894. *Materials for the Study of Variation.* London: Macmillan.

Cock, A. G. 1973. William Bateson, mendelism and biometry. *J. Hist. Biol.* 6:1–36.

DISLOCATION HYPOTHESIS *see* **NAVASHINIAN FUSION**

DISPERSAL THEORY [of the origin of insect flight]

Wigglesworth, V. B. 1963. Origin of flight in insects. *Proc. Roy. Entomol. Soc. Lond. C* 28:23–32.

——— . 1976. The evolution of insect flight. *Insect Flight.* Edited by R. C. Rainey. (*Symp. Roy. Entomol. Soc., Lond.* no. 7.) Oxford: Blackwell, pp. 255–69.

DISPLACEMENT THEORY [biogeography]

= CONTINENTAL DRIFT

Good, R. D'O. 1930. The geography of the genus *Coriaria. New Phytol.* 29:170–98.

DISRUPTIVE SELECTION [evolution]

= centrifugal selection = diversifying selection

Mather, K. 1953. The genetical structure of populations. *Symp. Soc. Exp. Biol.* 7:66–95.

——— . 1955. Polymorphism as an outcome of disruptive selection. *Evolution* 9:52–61.

Simpson, G. G. 1955. *The Major Features of Evolution*. New York: Columbia University Press, pp. 150, 153*ff*.

Thoday, J. M., T. B. Boam, & E. Millicent. 1959–61. Effects of disruptive selection. I-IV. *Heredity* 13:187–203 (Thoday); 205–18 (Thoday & Boam); 14:35–49 (Thoday); 16:199–217 (Millicent & Thoday).

Maynard Smith, J. 1962. Disruptive selection, polymorphism, and sympatric speciation. *Nature* 195:60–62.

Thoday, J. M. 1972. Disruptive selection. *Proc. Roy. Soc. Lond. B* 182:109–43.

Cf. FREQUENCY-DEPENDENT SELECTION; STABILIZING SELECTION

DISSOCIATION [of chromosomes]

White, M. J. D. 1957. Cytogenetics of the grasshopper *Moraba scurra*. I. Meiosis of interracial and interpopulation hybrids. *Aust. J. Zool.* 5:285–304.

______ . 1973. *Animal Cytology and Evolution*. 3d ed. Cambridge: Cambridge University Press.

Cf. ROBERTSONIAN CHANGES

DISTRIBUTION THEORY OF SEGMENT LENGTHS [in chromosomes]

Wilks, S. S. 1962. *Mathematical Statistics*. New York: Wiley, pp. 237–38.

Franklin, I. R. 1977. The distribution of the proportion of the genome which is homozygous by descent in inbred individuals. *Theor. Pop. Biol.* 11:60–80.

DIVERGENCE INDEX *see* **WAGNER DIVERGENCE INDEX**

DIVERGENCE RULE *see* **EICHLER'S RULE**

DIVERGENT DICHOTOMOUS MUTATION THEORY

Willis, J. C. 1949. *The Birth and Spread of Plants*. Waltham, Mass.: Chronica Botanica, chap. 8*ff*.

Good, R. 1974. *The Geography of the Flowering Plants*. 4th ed. London: Longman, p. 41.

DIVERGENT EVOLUTION

Gulick, J. T. 1878. Divergent evolution through cumulative segregation. *J. Linn. Soc. Lond. Zool.* 20:189–274.

Lesch, J. E. 1975. The role of isolation in evolution: George J. Romanes and John T. Gulick. *Isis* 66:483-503.

Cf. GEOGRAPHIC ISOLATION; REPRODUCTION ISOLATION

DIVERSIFYING SELECTION *see* **DISRUPTIVE SELECTION**

DIVERSITY [ecology; evolution]

MacArthur, R. H. 1965. Patterns of species diversity. *Biol. Rev.* 40:510–33.

Pianka, E. R. 1966. Latitudinal gradients in species diversity: a review of concepts. *Amer. Nat.* 100:33-46.

Diversity and Stability in Ecological Systems. 1969. *Brookh. Symp. Biol.*, no. 22.

Johnson, M. P., & P. H. Raven. 1970. Natural regulation of plant species diversity. *Evol. Biol.* 4:127–62.

Hendrickson, J. A., Jr., & P. R. Ehrlich. 1971. An expanded concept of "species diversity." *Not. Nat.* no. 239.

Hurlbert, S. H. 1971. The nonconcept of species diversity: a critique and alternative parameters. *Ecology* 52:577-86.

Fager, E. W. 1972. Diversity; a sampling study. *Amer. Nat.* 106:293–309.

Whittaker, R. H. 1972. Evolution and measurement of species diversity. *Taxon* 21:213–51.

______ . 1977. Evolution of species diversity in land communities. *Evol. Biol.* 10:1–67.

Cf. SHANNON-WEAVER DIVERSITY INDEX

DIVERSITY-STABILITY THEORY or HYPOTHESIS [ecology]

Hairston, N. G., J. O. Allan, R. K. Colwell, D. J. Futuyma, J. Howell, M. D. Lubin, J. Mathias, & J. H. Vandermeer. 1968. The relationship between species diversity and stability: an experimental approach with protozoa and bacteria. *Ecology* 49:1091–101.

Lewontin, R. C. 1969. The meaning of stability. *Diversity and Stability in Ecological Systems. Brookh. Symp. Biol.* 22:13–24.

Murdoch, W. W., F. C. Evans, & C. H. Peterson. 1972. Diversity and pattern in plants and insects. *Ecology* 53:819–29.

May. R. M. 1973. *Stability and Complexity in Model Ecosystems*. Princeton, N.J.: Princeton University Press.

Goodman, D. 1974. The validity of the diversity-stability hypothesis. *Structure, Functioning and Management of Ecosystems*. First International Congress of Ecology, The Hague, Sept. 8-14, 1974. Wageningen: Centre for Agricultural Publishing and Documentation, pp. 75–79.

______ . 1975. The theory of diversity-stability relationships in ecology. *Q. Rev. Biol.* 50:237–66.

Orians, G. H. 1975. Diversity, stability and maturity in natural ecosystems. *Unifying Concepts in Ecology*. Edited by W. H. Van Dobben & R. H. Lowe-McConnell. The Hague: Junk, pp. 139–50.

Cf. MAY'S PARADOX

DIXON-ASKENAZY COHESION THEORY *see* **COHESION THEORY**

DIXON'S COHESION THEORY *see* **COHESION THEORY**

DNA CONSTANCY RULE or HYPOTHESIS [genetics]

Ris, H., & A. E. Mirsky. 1949. Quantitative cytochemical determination of deoxyribonucleic acid with the Feulgen nucleal reaction. *J. Gen. Physiol.* 33:125–45.

Swift, H. 1950. The constancy of desoxyribose nucleic acid in plant nuclei. *Proc. Natl. Acad. Sci.* 36:643–54.

______. 1962. Nucleic acids and cell morphology in dipteran salivary glands. *The Molecular Control of Cellular Activity*. Edited by J. M. Allen. New York: McGraw-Hill, pp. 73–125.

Thomas, C. A., Jr. 1971. The genetic organization of chromosomes. *Annu. Rev. Genet.* 5:237–56.

DNA HOMOLOGY [genetics]

McCarthy, B. J., & E. T. Bolton. 1963. An approach to the measurement of genetic relatedness among organisms. *Proc. Natl. Acad. Sci.* 54:1636–41.

de Ley, J. 1968. Molecular biology and bacterial phylogeny. *Evol. Biol.* 2:103–56.

Cf. HOMOLOGY

DOLLO'S LAW [irreversibility in evolution]

Dollo, L. 1893. Les lois de l'evolution. *Bull. Soc. Belge. Geol. Paleont. Hydrob.* 7:164–66.

______. 1922. Les céphalopodes déroulés et l'irreversibilité de l'évolution. *Bijdr. Dierk.* 22:215–26.

Gregory, W. K. 1936. On the meaning and limits of irreversibility in evolution. *Amer. Nat.* 70:517–28.

Cave, A. J. E. & R. W. Haines. 1944. Meristic variation and reversibility of evolution. *Nature* 154:579–80.

Remane, A. 1952. *Die Grundlagen des natürlichen Systems, der Vergleichenden Anatomie und der Phylogenetik. Theoretische Morphologie und Systematik.* I. Leipzig: Akad. Verlags. Geest & Portig.

Farris, J. S. 1977. Phylogenetic analysis under Dollo's law. *Syst. Zool.* 26:77–88.

Simpson, G. G. 1953. *The Major Features of Evolution*. New York: Columbia University Press, pp. 310–11.

Wiedmann, J. 1969. The heteromorphs and ammonoid extinction. *Biol. Rev.* 44:563–602.

Gould, S. J. 1970. Dollo on Dollo's law: irreversibility and the status of evolutionary laws. *J. Hist. Biol.* 3:189–212.

Cf. IRREVERSIBILITY, LAW OF; UNIQUELY EVOLVED CHARACTER CONCEPT

DOMAIN HYPOTHESIS [immunology]

Edelman, G. M. 1971. Antibody structure and molecular immunology. *Ann. N.Y. Acad. Sci.* 190:5–25.

DOMINANCE [behavior]

Morse, D. H. 1974. Niche breadth as a function of social dominance. *Amer. Nat.* 108:818–30.

Rowell, T. E. 1974. The concept of dominance. *Behav. Biol.* 11:131–54.

Wilson, E. O. 1975. *Sociobiology*. Cambridge: Belknap Press of Harvard University Press, chap. 13.

DOMINANCE [ecology]

Clements, F. E. 1928. *Plant Succession and Indicators*. New York: H. W. Wilson, p. 72.

Whittaker, R. H. 1958. Recent evolution of ecological concepts in relation to the eastern forests of North America. *Fifty Years of Botany*. Edited by W. C. Steere. New York: McGraw-Hill, pp. 340–58.

DOMINANCE [genetics]

Fisher, R. A. 1922. On the dominance ratio. *Proc. Roy. Soc. Edinb.* 42:321–41.

Ford, E. B. 1930. The theory of dominance. *Amer. Nat.* 64:560–65.

Cf. FISHER'S THEORY OF DOMINANCE; MENDEL'S LAWS

DOMINANCE HYPOTHESIS [of hybrid vigor]

Bruce, A. B. 1910. The Mendelian theory of heredity and the augmentation of vigor. *Science* 32:627–28.

Keeble, F. & C. Pellew. 1910. The mode of inheritance of stature and flowering time in peas *(Pisium sativarn)*. *J. Genet.* 1:47–56.

Crow, J. R. 1948. Alternative hypotheses of hybrid vigor. *Genetics* 33:477–87.

Cf. HETEROSIS

DOSAGE EFFECT [genetics]

= dosage compensating effect

Stern, C. 1929. Uber die additive Wirkung multipler Allele. *Biol. Zblt.* 49:261–90.

______. 1960. Dosage compensation—development of a concept and new facts. *Can. J. Genet. Cytol.* 2:105–18.

Muller, H. J. 1932. Further studies on the nature and causes of gene mutations. *Proc. 6th Int. Cong. Genet.* 1:213–55.

Lucchesi, J. C. 1973. Dosage compensation in *Drosophila. Annu. Rev. Genet.* 7:225–37.

DOWN-GRADE THEORY [of bryophyte evolution]

Richards, P. W. 1959. Bryophyta. *Vistas in Botany*. Edited by W. B. Turrill. New York: Pergamon, I:387–420.

DRIESCH'S MACHINE THEORY OF LIFE

Driesch, H. 1894. *Analytische Theorie der Organischen Entwicklung*. Leipzig: Engelmann.

Churchill, F. B. 1969. From machine-theory to entelechy: two studies in development teleology. *J. Hist. Biol.* 2:165–85.

DRIFT *see* **GENETIC DRIFT; SEWALL WRIGHT EFFECT**

DRIFT SEQUENCE RULE [biogeography; systematics]

Brundin, L. 1965. On the real nature of transantarctic relationships. *Evolution* 19:496–505.

Edmunds, G. F., Jr. 1972. Biogeography and evolution of Ephemeroptera. *Annu. Rev. Entomol.* 17:21–42.

Ashlock, P. D. 1974. The uses of cladistics. *Annu. Rev. Ecol. Syst.* 5:81–99.

DRIVE, CONCEPT OF [behavior]

Hinde, R. A. 1956. Ethological models and the concept of drive. *Brit. J. Philos. Sci.* 6:321–33.

______. 1959. Unitary drives. *Anim. Behav.* 7:130–41.

______. 1970. *Animal Behavior: A Synthesis of Ethology and Comparative Psychology*. 2d ed. New York: McGraw-Hill, pp. 193*ff*.

Cicala, G. A. 1965. *Animal Drives*. New York: Von Nostrand.

Bolles, R. C. 1967. *Theory of Motivation*. New York: Harper & Row.

Cofer, C. N., & M. H. Appley. 1964. *Motivation. Theory and Research*. New York: Wiley.

Cf. MOTIVATION

DUAL-MECHANISM HYPOTHESIS [origin of complete mutations]

Reissig, J. L. 1963. Induction of forward mutants in the pyr-3 region of *Neurospora*. *J. Gen. Microbiol*. 30:317–25.

Nasim, A., & C. Auerbach. 1967. The origin of complete and mosaic mutants from mutagenic treatment of single cells. *Mutat. Res*. 4:1–14.

Auerbach, C., & B. J. Kilbey, 1971. Mutation in eukaryotes. *Annu. Rev. Genet*. 5:163–218.

Cf. LETHAL-HIT HYPOTHESIS; MASTER-STRAND HYPOTHESIS; REPAIR HYPOTHESIS

DUBININ EFFECT [genetics]

Dubinin, N. P., & B. N. Sidorov. 1934. Relation between the effect of a gene and its position in the system. *Amer. Nat*. 68:377–81.

Lewis, E. B. 1950. The phenomenon of position effect. *Adv. Genet*. 3:73–115.

Stern, C., & M. Kodani. 1955. Studies on the position effect at the cubitus interruptus locus of *Drosophila melanogaster*. *Genetics* 40:343–73.

Cf. POSITION EFFECT

DUPLICATION-DIFFERENTIATION CYCLE [of increasing genetic complexity]

Bridges, C. B. 1935. Salivary chromosome maps. *J. Hered*. 26:60–64.

Ohno, S. 1970. *Evolution by Gene Duplication*. Berlin: Springer-Verlag, part 3.

Stebbins, G. L. 1976. Chromosome, DNA and plant evolution. *Evol. Biol*. 9:1–34.

DURIAN THEORY [evolution]

Corner, E. J. H. 1949. The Durian theory or the origin of the modern tree. *Ann. Bot*. 13:367–414.

——— . 1953–4. The Durian theory extended. I-III. *Phytomorphology* 3:465–76; 4:152–65, 263–74.

DYAR'S RULE or LAW [of growth ratio to head size]

= Brook's law

Brooks, W. K. 1886. *Report on the Stomatapoda Collected by H.M.S. Challenger during the Years 1873–1876*. Challenger Zoological Reports, pt. 45.

Dyar, H. G. 1890. Number of moults in lepidopterous larvae. *Psyche* 5:420–22.

Taylor, R. L. 1931. On "Dyar's Rule" and its application to sawfly larvae. *Ann. Entomol. Soc. Amer.* 24:451–66.

Peters, W. 1972. Analysis of growth phases in *Krizousacorixa femorata* (Guerin) (Heteroptera). *Trans. Conn. Acad. Arts Sci.* 44:285–90.

DYNAMIC EQUILIBRIUM HYPOTHESIS [evolution]

Bigelow, R. S. 1965. Hybrid zones and reproductive isolation. *Evolution* 19:449–58.

Moore, W. S. 1977. An evaluation of narrow hybrid zones in vertebrates. *Q. Rev. Biol.* 52:263–77.

Huston, M. 1979. A general hypothesis of species diversity. *Amer. Nat.* 113:81–101.

DZIERZON'S THEORY [honeybee sex determination]

Dzierzon, J. 1849. *Neue Verbesserte Bienen-zucht des Pfarners Dzierzon*. Neisse: Wangenfield.

Berlepsch, A. von. 1884. *The Dzierzon Theory*. Medina, Ohio: Root.

Folsom, J. W. 1922. *Entomology with Special Reference to Its Ecological Aspects*. Philadelphia: Blakiston, pp. 286–87.

[E]

"EAT AND RUN" HYPOTHESIS [feeding behavior; evolution]

Young, J. Z. 1950. *The Life of Vertebrates*. New York: Oxford University Press, chap. 31.

Colbert, E. H. 1955. *The Evolution of the Vertebrates*. New York: Wiley, chap. 28.

Janis, C. 1976. The evolutionary strategy of the Equidae and the origins of rumen and cecal digestion. *Evolution* 30:757–74.

ECHOLOCATION *see* **BEAT THEORY**

ECOGEOGRAPHIC CHARACTER VARIATION [ecology; evolution]

Van Valen, L. 1965. Morphological variation and width of ecological niche. *Amer. Nat.* 99:377–90.

Brown, J. H., & A. K. Lee. 1969. Bergmann's rule and climatic adaptation in woodrats *(Neotoma)*. *Evolution* 23:329–38.

Snyder, G. K., & W. W. Weathers. 1975. Temperature adaptations in amphibians. *Amer. Nat.* 109:93–101.

ECOGEOGRAPHICAL RULES *see* **ALLEN'S RULE; BERGMANN'S RULE; WING RULE**

ECOLOGICAL DEATH

= density-dependent mortality

Lack, D. 1954. *The Natural Regulation of Animal Numbers.* London: Oxford University Press, chap. 11.

Bonner, J. T. 1965. *Size and Cycle.* Princeton, N.J.: Princeton University Press, p. 170.

ECOLOGICAL EFFICIENCY

= biological efficiency = efficiency

Lindeman, R. L. 1942. The trophic-dynamic aspect of ecology. *Ecology* 23:399–418.

Odum, E. P. 1971. *Fundamentals of Ecology.* 3d ed. Philadelphia: Saunders, p. 75.

Slobodkin, L. B. 1960. Energy relationships at the population level. *Amer. Nat.* 94:213–36.

______ . 1972. On the inconsistency of ecological efficiency and the form of ecological theories. *Trans. Conn. Acad. Arts Sci.* 44:291–305.

McClendon, J. H. 1975. Efficiency. *J. Theor. Biol.* 49:213–18.

Conrad, M. 1977. The thermodynamic meaning of ecological efficiency. *Amer. Nat.* 111:99–106.

ECOLOGICAL ISOLATION [biogeography; evolution]

= intrageneric isolation

Tretzel, E. 1955. Intragenerische Isolation und interspezifische Konkurrenz bei Spinnen. *Z. Morphol. Oekol. Tiere* 44:43–162.

Lack, D. L. 1971. *Ecological Isolation in Birds.* Cambridge: Harvard University Press.

Levin, D. A. 1978. The origin of isolating mechanisms in flowering plants. *Evol. Biol.* 11:185–317.

ECOLOGICAL NICHE *see* **NICHE**

ECOLOGICAL RACE [genetics; systematics]

Clausen, J. D., D. Keck, & W. M. Hiesey. 1940. Experimental studies on the nature of species. I. Effect of varied environments on western North American plants. *Carnegie Inst. Wash. Publ.* no. 520.

Clausen, J. 1953. The ecological race as a variable biotype compound in dynamic balance with its environment. *I. U.*

B. S. *Symposium on Genetics of Population Structure.* Pavia, Italy. Pp. 105–13.

——— . , & W. M. Hiesey. 1958. Experimental studies on the nature of species. IV. Genetic structure of ecological races. *Carnegie Inst. Wash. Publ.* no. 615.

Grant, V. 1954. Genetic and taxonomic studies in *Gilia.* IV. *Gilia achilleaefolia. Aliso* 3:1–18.

——— . 1977. *Organismic Evolution.* San Francisco: Freeman, p. 168.

Böcher, T. W. 1977. Convergence as an evolutionary process. *Bot. J. Linn. Soc.* 75:1–19.

ECOLOGICAL RECIPROCITY [ecology]

Turcek, F. J. 1963. Color preference in fruit- and seed-eating birds. *Proc. XII Int. Ornithol. Congr.* I:285–92.

ECOLOGICAL RELEASE *see* **NICHE EXPANSION**

ECOLOGICAL STABILITY *see* **STABILITY, ECOLOGICAL**

ECONOMY IN NATURE *see* **OPTIMALITY; PARSIMONY**

ECONOMY OF GROWTH ENERGY *see* **MATERIAL COMPENSATION**

ECOSPACE [ecology]

Valentine, J. W. 1969. Patterns of taxonomic and ecological structure of the shelf benthos during Phanerozoic time. *Palaeontology* 12:684–709.

Cf. BIOSPACE

ECOSPECIES [systematics; ecology]

Turesson, G. 1922. The genotypical response of the plant species to the habitat. *Hereditas* 3:211–350.

Axelrod, D. I. 1941. The concept of ecospecies in Tertiary paleobotany. *Proc. Natl. Acad. Sci.* 27:545–51.

Valentine, D. H. 1949. The units of experimental taxonomy *Acta Biotheor.* 975–88.

Cain, A. J. 1953. Geography, ecology and coexistence in relation to the biological definition of the species. *Evolution* 7:76–83.

ECOSYSTEM [ecology]

Tansley, A. G. 1935. Use and abuse of vegetational concepts and terms. *Ecology* 16:284–307.

Evans, F. G. 1956. Ecosystem as the basic unit in ecology. *Science* 123:1127–28.

Gates, D. M. 1968. Toward understanding ecosystems. *Adv. Ecol. Res.* 5:1–35.

Odum, E. P. 1969. The strategy of ecosystem development. *Science* 164:262–70.

ECOSYSTEM CONVERGENCE [ecology]

Cody, M. L. 1974. Optimization in ecology. *Science* 183:1156–64.

Mooney, H. A. 1974. Plant forms in relation to environment. *Vegetation and Environment*. Edited by B. R. Strain & W. D. Billings. (*Handbook of Vegetation Science*, no. 6.) The Hague: Junk, pp. 111–22.

Peet, R. K. 1978. Ecosystem convergence. *Amer. Nat.* 112:441–44.

ECOTONE [ecology]

Allee, W. C., A. E. Emerson, O. Park, T. Park, & K. P. Schmidt. 1949. *Principles of Animal Ecology*. Philadelphia: Saunders, p. 476.

Clements, F. E. 1905. *Research Methods in Ecology*. Lincoln, Neb.: University Publishing.

Cf. TENSION ZONE

ECOTOPE [ecology]

Schmithüsen, J. 1968. Allgemeine Vegetationsgeographie. *Lehrbuch der allgemeinen Geographie*. Edited by E. Obst & J. Schmithüsen. Berlin: DeGruyter. 3d ed. Part 4.

Whittaker, R. H., S. A. Levin, & R. B. Rost. 1973. Niche, habitat, and ecotope. *Amer. Nat.* 107:321–38.

ECOTYPE [systematics; ecology]

Turesson, G. 1922. The species and the variety as ecological units. *Hereditas* 3:100–13.

———. 1925. The plant species in relation to habitat and climate. *Hereditas* 4:147–236.

Gregor, J. W. 1944. The ecotype. *Biol. Rev.* 19:20–30.

Turrill, W. B. 1946. The ecotype concept: a consideration with appreciation and criticism, especially of recent trends. *New Phytol.* 45:34–43.

Bocher, T. W. 1977. Convergence as an evolutionary process. *Bot. J. Linn. Soc.* 75:1–19.

Valentine, D. H. 1978. Ecological criteria in plant taxonomy. *Essays in Plant Taxonomy*. Edited by H. E. Street. London & New York: Academic Press, pp. 1–18.

EDDY DIFFUSION THEORY [pollination biology]

Sutton, O. G. 1932. A theory of eddy diffusion in the atmosphere. *Proc. Roy. Soc. Lond.* A 135:143–65.

Levin, D. A., & H. W. Kerster. 1974. Gene flow in seed plants. *Evol. Biol.* 7:139–220.

EFFICIENCY *see* **ECOLOGICAL EFFICIENCY**

EFFORT, LAW OF [behavior]

Hess, E. H. 1958. "Imprinting" in animals. *Sci. Amer.* 198(3):81–90.

______. 1959. Imprinting. *Science* 130:133–41.

EHRLICH'SHCE PHÄNOMEN [immunology]

= Ehrlich's phenomenon = le phenomène de Danysz

Ehrlich, P. 1897. Die Wertbemessung des Diptherieheilserums, und deren theoretische Grundlagen. *Klin. Jhb.* 6:299–366. Reprinted in: *Collected Papers of Paul Ehrlich.* Edited by F. Himmelweit. New York: Pergamon, 1960. II:86–106.

Mazumdar, P. M. H. 1974. The antigen-antibody reaction and the physics and chemistry of life. *Bull. Hist. Med.* 48:1–21.

EHRLICH'S SIDE-CHAIN THEORY [immunology]

= receptory theory = side-chain theory

Ehrlich, P. 1885. *The Requirement of the Organs for Oxygen.* Berlin: A. Hirschwald. Engl. transl. in: *Collected Papers of Paul Ehrlich.* Edited by F. Himmelweit. New York: Pergamon, 1956. I:433–96.

______. 1897. Die Wertbemessung des Diptherie-heilserums, und deren theoretische Grundlagen. *Klin. Jhb.* 6:299–326. Reprinted in: *Collected Papers of Paul Ehrlich.* Edited by F. Himmelweit, New York: Pergamon, 1960. II:86–106.

______. 1900. On immunity with special reference to cell life. *Proc. Roy. Soc. Lond.* 66:424–48.

Heymann, B. 1928. Zur Geschichte der Seitenkettentheorie Paul Ehrlichs. *Klin. Wochenschr.* 7:1257–305.

Bulloch, W. 1938. *The History of Bacteriology.* London: Oxford University Press, chap. 11.

Parascandola, J., & R. Jasensky. 1974. Origins of the receptor theory of drug action. *Bull. Hist. Med.* 48:199–220.

EICHLER'S RULE [parasitology]

= divergence rule

Eichler, W. 1948. Some rules in ectoparasitism. *Ann. Mag. Nat. Hist.*, ser. 12, 1:588–98.

Noble, E. R., & G. A. Noble. 1964. *Parasitology*. 2d ed. Philadelphia: Lea & Febiger, chap. 26.

ELECTRICAL TRANSMISSION THEORY [of intercellular communication]

du Bois-Reymond, E. 1877. Experimentalkritik der Entladungshypothese über die Wirkung von Nerf auf Muskel. *Gesammelte Abhandlungen zur allgemeinen Muskel-und Nervenphysik*. Vol. II. Leipzig.

Grundfelt, H. 1975. History of the synapse as a morphological and functional structure. *Golgi Centennial Symposium*. Edited by M. Santini. New York: Raven, pp. 39–50.

ELECTRO-DYNAMIC THEORY [of life]

Northrop, F. S. C., & H. S. Burr. 1937. Experimental findings concerning the electro-dynamic theory of life and an analysis of their physical meaning. *Growth* 1:78–88.

Wardlaw, C. W. 1952. *Phylogeny and Morphogenesis*. London: Macmillan, p. 456.

ELTONIAN PYRAMID [food chain in an ecosystem]

= food pyramid

Elton, C. 1927. *Animal Ecology*. New York: Macmillan.

Lindeman, R. L. 1942. The trophic-dynamic aspect of ecology. *Ecology* 23:399–418.

EMBOITEMENT THEORY *see* **PREFORMATION THEORY**

EMERGENCE [evolution]

= emergent evolution

Morgan, C. L. 1923. *Emergent Evolution*. Gifford Lectures, 1922. London.

———. 1927. *Emergent Evolution*. New York: Holt.

Schievella, P. S. 1973. Emergent evolution and reductionism. *Scientia* 108:323–30.

EMERY'S RULE [relation among hymenopterous insect parasites]

Emery, C. 1909. Über den Ursprung der dulotischen, parasitischen und myrmekophilen Ameisen. *Biol. Zblt.* 29:352–62.

Le Masne, G. 1956. Recherches sur les fourmis parasites *Plagiolepis grassei* et l'evolution des *Plagiolepis* parasites. *C. R. Acad. Sci.* 243:673–75.

Wilson, E. O. 1971. *The Insect Societies*. Cambridge: Harvard University Press, chap. 19.

EMPHATIC LEARNING [behavior]

Klopfer, P. H. 1957. An experiment on emphatic learning in ducks. *Amer. Nat.* 91:61–63.

______ . 1959. Social interactions in discrimination learning with special reference to feeding behavior in birds. *Behaviour.* 14:282–99.

ENDOCYCLIC SELECTION

= cyclical selection

McWhirter, K. G. 1967. Quantum-genetics of human blood-groups and phoneme-preferences. *Heredity* 22:162–63.

Pasteur, G. 1977. Endocyclic selection in reptiles. *Amer. Nat.* 111:1027–30.

ENDOGENOUS TIMER HYPOTHESIS [of biological clocks; physiology]

Brown, F. A. 1968. "Endogenous" biorhythmicity reviewed with new evidence. *Scientia* 103:245–60.

Laar, W. van. 1970. A contribution to the problem of the concept "biological clock." Part I. *Acta Biotheor.* 19:95–139.

Cf. BIOLOGICAL CLOCK

ENDOMIXIS [morphogenesis]

Woodruff, L. L., & R. Erdmann. 1914. A normal periodic reorganization process without cell fusion in Paramecium. *J. Exp. Zool.* 17:425–520.

Sonneborn, T. M. 1954. The relation of autogamy to senescence and rejuvenescence in *Paramecium aurelia*. *J. Protozool.* 1:38–53.

ENDOSYMBIOTIC HYPOTHESIS [of the origin of eukaryotic cells] *see* **SYMBIOTIC THEORY**

ENERGY-ALLOCATION PRINCIPLE [evolution]

Cole, L. C. 1954. The population consequences of life history phenomena. *Q. Rev. Biol.* 29:103–37.

Gadgil, M., & W. H. Bossert. 1970. Life historical consequences of natural selection. *Amer. Nat.* 104:1–24.

Demetrius, L. 1975. Reproductive strategies and natural selection. *Amer. Nat.* 109:243–49.

Cf. r AND K SELECTION

ENERGY COUPLING [in mitochondria]

Green, D. E., & G. A. Blondin. 1978. Molecular mechanism of mitochondrial energy coupling. *Bioscience* 28:18–24.

ENERGY FLOW [in ecosystems]

Juday, C. 1940. The annual energy budget of an inland lake. *Ecology* 21:438–50.

Odum, E. P. 1968. Energy flow in ecosystems: a historical review. *Amer. Zool.* 8:11–18.

Mann, K. H. 1969. The dynamics of aquatic ecosystems. *Adv. Ecol. Res.* 6:1–81.

Rigler, F. H. 1975. The concept of energy flow and nutrient flow between trophic levels. *Unifying Concepts in Ecology*. Edited by W. H. van Dobben & R. H. Lowe-McConnell. The Hague: Junk, pp. 15–26.

Cf. TROPHIC-DYNAMIC CONCEPT

ENERGY HYPOTHESIS [of primate feeding behavior]

Hamilton, W. J., III, & C. D. Busse. 1978. Primate carnivory and its significance to human diets. *Bioscience* 28:761–66.

ENTELECHY [evolution; philosophy]

Driesch, H. 1903. Kritisches und Polemisches. IV. Zur Verständigung über die "Entelechie." *Biol. Zblt.* 23:697–704.

Radl, E. 1930. *The History of Biological Theories*. Transl. by E. J. Hatfield. London: Oxford University Press, chap. 32.

Simpson, G. G. 1953. *The Major Features of Evolution*. New York: Columbia University Press, p. 133.

Churchill, F. B. 1969. From machine-theory to entelechy: two studies in developmental teleology. *J. Hist. Biol.* 2:165–85.

Cf. ARISTOGENESIS

ENTEROCOEL THEORY [of protozoan origin; evolution]

Hartman, W. D. 1963. A critique of the enterocele theory. *The Lower Metazoa*. Edited by E. C. Dougherty. Berkeley: University of California Press, pp. 55–77.

ENTROPY, LAW OF [evolution; development]

Bertalanffy, L. von. 1950. The theory of open systems in physics and biology. *Science* 111:23–29.

Nowinski, W. W. 1960. *Fundamental Aspects of Normal and Malignant Growth*. Amsterdam: Elsevier, p. 155.

Walker, I. 1976. Maxwell's demon in biological systems. *Acta Biotheor.* 25:103–110.

Jones, D. D. 1977. Entropic models in biology:the next scientific revolution? *Perspect. Biol. Med.* 20:285–99.

ENVIRONMENTAL-GRAIN VARIATION THEORY [of the niche; ecology]

Ludwig, W. 1950. Zur Theorie der Konkurenz. Die Annidation (Einnischung) als fünfter Evolutionsfaktor. *Neue Ergebnisse und Probleme Zoologie*. Edited by W. Herre. Leipzig: Geest & Portig, pp. 516–37.

Levene, H. 1953. Genetic equilibrium when more than one ecological niche is available. *Amer. Nat.* 87:331–33.

Soule, M. 1976. Allozyme variation: its determinants in space and time. *Molecular Evolution*. Edited by F. J. Ayala. Sunderland, Mass.: Sinauer Associates, pp. 60–77.

ENVIRONMENTAL HETEROGENEITY [ecology; evolution]

Hedrick, P. W., M. E. Ginevan, & E. P . Ewing. 1976. Genetic polymorphism in heterogeneous environments. *Annu. Rev. Ecol. Syst.* 7:1–32.

Spieth, P. T. 1979. Environmental heterogeneity: a problem of contradictory selection pressures, gene flow, and local polymorphism. *Amer. Nat.* 113:247–60.

ENVIRONMENTAL HETEROGENEITY HYPOTHESIS [of species diversity]

Paine, R. T. 1966. Food web complexity and species diversity. *Amer. Nat.* 100:65–75.

Huston, M. 1979. A general hypothesis of species diversity. *Amer. Nat.* 113:81–101.

ENVIRONMENTAL PATCHINESS *see* **PATCHY ENVIRONMENT**

ENZYME THEORY [of life]

Hofmeister, F. 1901. *Die chemische Organization der Zelle*. Braunschweig: Viewig, p. 14.

Kohler, R. E. 1972. The reception of Eduard Buchner's discovery of cell-free fermentation. *J. Hist. Biol.* 5:327–53.

______ . 1973. The enzyme theory and the origins of biochemistry. *Isis* 64:181–96.

Cf. PROTOPLASM THEORY; ZYMASE HYPOTHESIS

EPAMINONDAS EFFECT [genetic adaptation]

Levins, R. 1961. Mendelian species as adaptive systems. *Gen. Syst.* 6:33–39.

EPHEMERAL-ZONE HYPOTHESIS [of vertebrate hybridization]

Dobzhansky, T. 1940. Speciation as a stage in evolutionary divergence. *Amer. Nat.* 74:312–21.

Moore, W. S. 1977. An evaluation of narrow hybrid zones in vertebrates. *Q. Rev. Biol.* 52:263–77.

EPIGENESIS, THEORY OF

Harvey, W. 1651. *Exercitationes de generatione animalium.* Amsterdam: Ravesteynium. (reprinted in: *The Works of William Harvey.* Translated by R. Willis. London: Sydenham Society 1847).

Cole, F. J. 1930. *Early Theories of Sexual Generation.* Oxford: Clarendon Press, chap. 6.

Needham, J. 1931. *Chemical Embryology.* Cambridge: Cambridge University Press.

Temkin, O. 1950. German concepts of ontogeny and history around 1800. *Bull. Hist. Med.* 24:227–46.

Waddington, C. H. 1953. Epigenetics and evolution. *Symp. Soc. Exp. Biol.* 7:186–99.

Roe, S. A. 1979. Rationalism and embryology: Casper Friedrich Wolff's theory of epigenesis. *J. Hist. Biol.* 12:1–43.

Cf. BIOGENESIS; ONTOGENESIS

EPIGENETIC HOMEOSTASIS

Nanney, D. L. 1958. Epigenetic control systems. *Proc. Natl. Acad. Sci.* 44:712–17.

Cf. CANALIZATION; DEVELOPMENTAL HOMEOSTASIS; HOMEOSTASIS

EPIGENETICS *see* **EPIGENESIS**

EPISODES OF PROLIFERATION *see* **EXPLOSIVE EVOLUTION**

EPISOMES [genetics]

Jacob, F., & E. L. Wollman. 1958. Les episomes, elements genetiques ajoutes. *C. R. Acad. Sci.* 247:154–56.

Campbell, A. M. 1962. Episomes. *Adv. Genet.* 11:101–45.

Scaife, J. 1967. Episomes. *Annu. Rev. Microbiol.* 21:601–38.

Silvestri, L. G., ed. 1973. *Possible episomes in eukaryotes. Proc. 4th Lepetit Colloq.*, 1972. Amsterdam: North Holland Publishing.

EPISTASIS CYCLE THEORY [population genetics]

Soule, M. 1973. The epistasis cycle: a theory of marginal populations. *Annu. Rev. Ecol. Syst.* 4:165–87.

EPITOKY

Bourgeois, A., & P. Cassagnau. 1973. Les perturbations morphogénétiques de type épitoque chez les Collemboles

Hypogastruridae. *C. R. Acad. Sci.* 277:1197–200.
Fjellberg, A. 1977. Etidoky [sic] in *Vertagopus* species (Collembola, Isotomidae). *Rev. Ecol. Biol. Sol* 14:493–95.

EQUABLE VARIABILITY, LAW OF [morphology; evolution]
Meyen, S. V. 1973. Plant morphology in its nomothetical aspects. *Bot. Rev.* 39:205–60.

EQUABLE VARIATIONS, LAW OF *see* **HOMOLOGOUS SERIES, LAW OF**

EQUAL OPPORTUNITY, PRINCIPLE OF [for species to colonize]
MacArthur, R. H. 1972. *Geographical Ecology*. New York: Harper & Row, pp. 173–76.

EQUATORIAL CONSTRICTION THEORY [of cell cleavage]
Ziegler, H. E. 1903. Experimentelle Studien über Zelltheilung. IV. Die Zelltheilung der Furchungzellen bei Beroë und Echinus. *Wilhelm Roux' Arch.* 16:155–75.
Rappaport, R. 1974. Cleavage. *Concepts of Development*. Edited by J. Lash & J. R. Whittaker. Stamford, Conn: Sinauer Associates, pp. 76-98.

EQUIFORMAL PROGRESSIVE AREAS, THEORY OF [biogeography]
= Hulten's hypothesis
Hulten, E. 1937. *Outline of the History of Arctic and Boreal Biota During the Quaternary Period*. Stockholm: Bokförlags akt. Thule.
Raup, H. M. 1941. Botanical problems in boreal America. I. *Bot. Rev.* 7:147–208.
______. 1947. Some natural floristic areas in boreal America. *Ecol. Monogr.* 17:221–34.
Good, R. 1974. *The Geography of the Flowering Plants*. 4th ed. London: Longmans, p. 192.

EQUILIBRIUM THEORY [of island biogeography]
= ISLAND BIOGEOGRAPHY, THEORY OF
Preston, F. W. 1962. The canonical distribution of commonness and rarity. *Ecology* 43:185–215, 410–32.
MacArthur, R. H., & E. O. Wilson. 1963. An equilibrium theory of insular zoogeography. *Evolution* 17:373–87.
______. 1967. *The Theory of Island Biogeography*. Princeton, N.J.: Princeton University Press.
Simberloff, D. S. 1974. Equilibrium theory of island biogeography and ecology. *Annu. Rev. Ecol. Syst.* 5:161–82.

———. 1976. Species turnover and equilibrium island biogeography. *Science* 194:572–78.
———. 1978. Colonization of islands by insects: immigration, extinction, and diversity. *Symp. Roy. Entomol. Soc. Lond.* 9:139–53.
Cf. NONEQUILIBRIUM HYPOTHESIS

EQUITABILITY *see* **FAVORABLENESS**

EQUIVALENCE RULE [in DNA Composition] *see* **CHARGAFF RULES**

EQUIVOCAL GENERATION *see* **SPONTANEOUS GENERATION**

ERGONOMICS [work, performance, and efficiency in insect societies]

Wilson, E. O. 1963. The social biology of ants. *Annu. Rev. Entomol.* 8:345–68.
———. 1968. The ergonomics of caste in the social insects. *Amer. Nat.* 102:41–66.

ERRERA'S LAW [cell division]

Errera, L. 1886. Sur une condition fondamentale d'équilibre des cellules vivantes. *C. R. Acad. Sci.* 103:822–24.
Thompson, D'A. W. 1942. *On Growth and Form.* 2d ed. Cambridge: Cambridge University Press, pp. 482–83.
Wardlaw, C. W. 1960. Some reflections on Errera's law. *Commemoration Leo Errera.* Universite libre de Bruxelles, 10–12 Sept., 1958. Bruxelles: Imprimerie Gutenberg, pp. 99–104.

ERUPTIVE EVOLUTION *see* **EXPLOSIVE EVOLUTION**

ESCAPE THEORY *see* **TRAP THEORY**

ESCAPE THEORY [of oncogenesis]

Needham, J. 1942. *Biochemistry and Morphogenesis.* Cambridge: Cambridge University Press, pp. 239*ff.*

ESSENTIALIST SPECIES CONCEPT *see* **SPECIES CONCEPTS**

ETHOLOGICAL ISOLATION

Wallace, A. R. 1890. *Darwinism: An Exposition of the Theory of Natural Selection.* London: Macmillan, chaps. 5–6.
Anderson, R. F. V. 1977. Ethological isolation and competition

of allospecies in secondary contact. *Amer. Nat.* 111:939–49.

ETHOLOGY

Jaynes, J. 1969. The historical origins of "ethology" and "comparative psychology." *Anim. Behav.* 17:601–06.

ETIOLOGY

Orlob, G. 1964. The concepts of etiology in the history of plant pathology. *Pflanz.-Nach. Bayer* 17:185–268.

EULER'S THEOREM [of cell surface pattern]

Euler, L. 1752–53. Elementa doctrinae solidorum. *Nov. Comm. Acad. Sci. Imp. Petrop.* 4:109–40.

Lewis, F. T. 1944. The geometry of growth and cell division in columnar parenchyma. *Amer. J. Bot.* 31:619–29.

EVOLUTION

Arber, A. 1907. Theories of evolution. An historical outline. *Naturalist, Hull:* 167–71, 209–15, 241–49.

Morgan, T. H. 1916. *A Critique of the Theory of Evolution.* Princeton, N.J.: Princeton University Press.

Glass, H. B., ed. 1959. *Forerunners of Darwin: 1745–1859.* Baltimore: Johns Hopkins Press.

Manser, A. R. 1965. The concept of evolution. *Philosophy* 40:18–34.

Simpson, G. G. 1969. The present status of the theory of evolution. *Proc. Roy. Soc. Vict.* 82:149–60.

Bowler, P. J. 1975. The changing meaning of "evolution." *J. Hist. Ideas* 36:95–114.

Cf. SPECIAL CREATION

EVOLUTIONAL LOAD *see* **SUBSTITUTIONAL LOAD**

EVOLUTIONARILY STABLE STRATEGY (ESS)

Maynard Smith, J. 1972. *On Evolution.* Edinburgh: Edinburgh University Press.

______ . 1974. The theory of games and the evolution of animal conflicts. *J. Theor. Biol.* 47:207–21.

______ , & G. A. Parker. 1976. The logic of asymmetric contests. *Anim. Behav.* 24:159–75.

Price, G, & J. Maynard Smith. 1973. The logic of animal conflict. *Nature* 246:15–18.

Lawlor, L. R., & J. Maynard Smith. 1967. The coevolution and stability of competing species. *Amer. Nat.* 110:79–99.

EVOLUTIONARY CANALIZATION

Stebbins, G. L. 1974. *Flowering Plants: Evolution above the*

Species Level. Cambridge: Belknap Press of Harvard University Press, p. 23.

EVOLUTIONARY CLOCK *see* **MOLECULAR EVOLUTIONARY CLOCK**

EVOLUTIONARY DANCE [cycles of adaptation and counteradaptation]

Whittaker, R. L. 1969. Evolution of species diversity in plant communities. *Brookh. Symp. Biol.* 22:178–85.

Levin, D. A. 1975. Pest pressure and recombination systems in plants. *Amer. Nat.* 109:437–51.

EVOLUTIONARY NOISE [population genetics]

Lewontin, R. C. 1974. *The Genetic Basis of Evolutionary Change.* New York: Columbia University Press, pp. 120, 246*ff.*

Brown, A. D. H. 1978. Isozymes, plant population genetic structure and genetic conservation. *Theor. Appl. Genet.* 52:145–57.

EVOLUTIONARY PARSIMONY *see* **PARSIMONY**

EVOLUTIONARY SPECIES [systematics]

Simpson, G. G. 1951. The species concept. *Evolution* 5:285–98.

——— . 1961. *Principles of Animal Taxonomy.* New York: Columbia University Press, chap. 5.

Cf. SPECIES CONCEPTS

EVOLUTIONARY SYSTEMATICS

Simpson, G. G. 1961. *Principles of Animal Taxonomy.* New York: Columbia University Press.

Mayr, E. 1969. *Principles of Systematic Zoology.* New York: McGraw-Hill.

Hull, D. L. 1970. Contemporary systematic philosophies. *Annu. Rev. Ecol. Syst.* 1:19–54.

Cf. PHENETIC SYSTEMATICS; PHYLOGENETIC SYSTEMATICS

EVOLUTIONARY TRENDS *see* **ORTHOGENESIS**

EXCHANGE HYPOTHESIS [of chromosome mutation]

Revell, S. H. 1959. The accurate estimation of chromatid breakage, and its relevance to a new interpretation of chromatid aberrations induced by ionizing radiations. *Proc. Roy. Soc. Lond. B* 150:563–89.

Evans, H. J. 1962. Chromosome aberrations induced by ionizing radiation. *Int. Rev. Cytol.* 13:221–321.

EXCLUSION *see* **COMPETITIVE EXCLUSION**

EXOGENOUS TIMER HYPOTHESIS [of biological clocks; physiology]

Brown, F. A. 1970. *The Biological Clock.* New York: Academic Press.

Laar, W. van. 1970. A contribution to the problem of the concept "biological clock." Part I. *Acta Biotheor.* 19:95–139.

Cf. BIOLOGICAL CLOCK

EXPLOSIVE EVOLUTION

Cloud, P. E., Jr. 1948. Some problems and patterns of evolution exemplified by fossil invertebrates. *Evolution* 2:322–50.

Simpson, G. G. 1953. *The Major Features of Evolution.* New York: Columbia University Press, p. 234.

Romer, A. S. 1960. Explosive evolution. *Zool. Jahrb. Abt. Syst.* 88:79–90.

EXTINCTION, LAW OF [evolution]

Brooks, J. L. 1972. Extinction and the origin of organic diversity. *Trans. Conn. Acad. Arts Sci.* 44:17–56.

MacArthur, R. H. 1972. *Geographical Ecology.* New York: Harper & Row.

Van Valen, L. 1973. A new evolutionary law. *Evol. Theory* 1:1–30.

Cf. RED QUEEN HYPOTHESIS

EYE OF A NEEDLE *see* **BOTTLENECK EFFECT**

[F]

FACTORIAL HYPOTHESIS [linear arrangement of genes]

Morgan, T. H. 1914. The mechanism of heredity as indicated by the inheritance of linked characters. *Pop. Sci. Mon.*, Jan., 1914, pp. 1–16.

Carlson, E. A. 1966. *The Gene: A Critical History.* Philadelphia: Saunders, pp. 30*ff.*

Cf. CHROMOSOME THEORY

FAHRENHOLZ'S RULE [parasite evolution]

Eichler, W. 1941. Wirtsspezifität und stammesgeschlichte

Gleichläufigkeit (Fahrenholzsche Regel) bei Parasiten in allgemeinen und bei Mallophagen im besonderen. *Zool. Anz.* 132:254–62.

———. 1948. Some rules in ectoparasitism. *Ann. Mag. Nat. Hist.*, ser. 12, 1:588–98.

Janiszewska, J. 1949. Parasitogenetic rules. Janicki rule. *Zool. Pol.* 5:31–34.

Henning, W. 1966. *Phylogenetic Systematics*. Urbana: University of Illinois Press, chap. 2.

FAMILIAR-AREA HYPOTHESIS [biogeography]

Baker, R. R. 1978. *The Evolutionary Ecology of Animal Migration*. New York: Holmes & Meier, chaps. 21–22, 32.

FAUNAL GATEWAY EFFECT [ecology; biogeography]

Cody, M. L. 1975. Towards a theory of continental species diversities. Bird distributions over Mediterranean habitat gradients. *Ecology and Evolution of Communities*. Edited by M. L. Cody & J. M. Diamond. Cambridge: Belknap Press of Harvard University Press, pp. 214–57.

FAVORABLENESS [ecology]

= equitability

Richards, P. W. 1952. *The Tropical Rain Forest*. Cambridge: Cambridge University Press.

Whittaker, R. H. 1965. Dominance and diversity in land plant communities. *Science* 147:250–60.

Monk, C. D. 1967. Tree species diversity in the eastern deciduous forest with particular reference to north central Florida. *Amer. Nat.* 101:173–87.

McNaughton, S. J., & L. L. Wolf. 1970. Dominance and the niche in ecological systems. *Science* 167:131–39.

Terborgh, J. 1973. On the notion of favorableness in plant ecology. *Amer. Nat.* 107:481–501.

FECHNER'S LAW *see* **ROSA'S RULE**

FEEDBACK CONTROL [ecology]

Wibbert, H. 1970. Cybernetic concepts in population dynamics. *Acta Biotheor.* 19:54–81.

Cf. GENETIC FEEDBACK

FEEDBACK THEORY [of evolution]

Riedl, R. 1977. A systems-analytical approach to macro-evolutionary phenomena. *Q. Rev. Biol.* 52:351–70.

FETALIZATION THEORY [human evolution]

= Bolk's fetalization theory

Bolk, L. 1926. *Das Problem der Menschwerdung*. Jena: Fischer.

Gould, S. J. 1977. *Ontogeny and Phylogeny*. Cambridge: Harvard University Press, pp. 356*ff*.

FIBONACCI PHYLLOTAXIS *see* **PHYLLOTAXIS**

FIELD CONCEPT [morphogenesis]

Gurwitsch, A. G. 1922. Über den Begriff des embryonalen Feldes. *Arch. Entw. Mech. Org.* 51:383–415.

______. 1944. *The Theory of Biological Field*. Moscow: Sov. Nauka (Russian).

Weiss, P. 1923. Die Regeneration der Urodelenextremität als Selbstdifferenzierung des Organrestes. *Naturwissenschaften* 11:669–77.

Huxley, J. S. 1935. The field concept in biology. *Trans. Dynam. Dev.* (Moscow). 10:269.

Needham, J. 1942. *Biochemistry and Morphogenesis*. Cambridge: Cambridge University Press, p. 127.

Wardlaw, C. W. 1952. *Phylogeny and Morphogenesis*. London: Macmillan, p. 448.

______. 1965. *Organization and Evolution in Plants*. London: Longmans, chap. 2.

Herrmann, H. 1964. Biological field phenomena: facts and concepts. *Form and Strategy in Science*. Edited by J. R. Gregg & F. T. C. Harris. Dordrecht, Netherlands: Reidel, pp. 343–62.

Waddington, C. H. 1966. Fields and gradients. *Major Problems in Developmental Biology*. Edited by M. Locke. New York: Academic Press, pp. 105–24.

Nieuwkoop, P. D. 1967. The "organization centre." II. Field phenomena, their origin and significance. III. Segregation and pattern formation in morphogenetic fields. *Acta Biotheor.* 17:151–77, 178–94.

Chandebois, R. 1976. Cell sociology: a way of reconsidering the current concepts of morphogenesis. *Acta Biotheor.* 25:71–102.

FIRST AVAILABLE SPACE THEORY [plant morphology]

Snow, M., & R. Snow. 1931–5. Experiments on phyllotaxis. I. The effect of isolating a primordium. II. The effect of displacing a primordium. III. Diagonal splits through decussate Apices. *Phil. Trans. Roy. Soc. Lond. B* 221:1–43; 222:353–400; 225:63–94.

———. 1948. On the determination of leaves. *Symp. Soc. Exp. Biol.* 2:263–75.

FISHER'S FUNDAMENTAL THEOREM [of natural selection]

Fisher, R. A. 1930. *The Genetical Theory of Natural Selection.* Oxford: Clarendon Press.

MacArthur, R. H. 1962. Some generalized theorems of natural selection. *Proc. Natl. Acad. Sci.* 48:1893–97.

Price, G. R. 1972. Fisher's "fundamental theorem" made clear. *Ann. Hum. Genet.* 36:129–40.

FISHER'S SEX RATIO THEORY [evolution]

= Fisher effect = Fisherian sexual selection

Fisher, R. A. 1930. *The Genetical Theory of Natural Selection.* Oxford: Clarendon Press.

Kolman, W. A. 1960. The mechanism of natural selection for the sex ratio. *Amer. Nat.* 94:373–77.

Trivers, R. L. 1972. Parental investment and sexual selection. *Sexual Selection and the Descent of Man.* Edited by B. Campbell. Chicago: Aldine, pp. 136–79.

Cf. HANDICAP PRINCIPLE; SEXUAL SELECTION

FISHER'S THEORY OF DOMINANCE [genetics]

Fisher, R. A. 1928. The possible modification of the response of the wild type to recurrent mutations. *Amer. Nat.* 62:115–26.

———. 1928. Two further notes on the origin of dominance. *Amer. Nat.* 62:571–74.

Wright, S. 1929. Fisher's theory of dominance. *Amer. Nat.* 63:274–79.

Haldane, J. B. S. 1930. A note on Fisher's theory of the origin of dominance and on a correlation between dominance and linkage. *Amer. Nat.* 64:87–90.

FITNESS [Darwinian]

Thoday, J. M. 1953. Components of fitness. *Symp. Soc. Exp. Biol.* 7:96–113.

Dobzhansky, T. 1968. Adaptedness and fitness. *Population Biology and Evolution.* Edited by R. C. Lewontin. Syracuse, N.Y.: Syracuse University Press, pp. 109–21.

———. 1968. On some fundamental concepts of Darwinian biology. *Evol. Biol.* 2:1–34.

Demetrius, L. 1977. Adaptedness and fitness. *Amer. Nat.* 111:1163–68.

Spiess, E. B. 1977. *Genes in Populations.* New York: Wiley, pp. 441*ff.*

FITNESS [genetics]

Levins, R. 1962. Theory of fitness in a heterogeneous environment. I. The fitness set and adaptive function. *Amer. Nat.* 96:361–78.

______ . 1965. Theory of fitness in a heterogeneous environment. V. Optimal genetic systems. *Genetics* 52:891–904.

Kojima, K. 1971. Is there a constant fitness value for a given genotype? No! *Evolution* 25:281–85.

FLIGHT THEORY [of feather evolution]

Parkes, K. 1966. Speculations of the origin of flight. *Living Bird* 5:77–86.

Regal, P. J. 1975. The evolutionary origin of feathers. *Q. Rev. Biol.* 50:35–65.

FLORAL ISOLATION, THEORY OF [evolution]

Grant, V. 1949. Pollination systems as isolating mechanisms in flowering plants. *Evolution* 3:82–97.

Straw, R. M. 1956. Floral isolation in *Penstemon*. *Amer. Nat.* 90:47–53.

Sprague, E. F. 1962. Pollination and evolution in *Pedicularis* (Scrophulariaceae). *Aliso* 5:181–209.

Grant, K. A., & V. Grant. 1964. Mechanical isolation of *Salvia apiana* and *Salvia mellifera* (Labiatae). *Evolution* 18:196–212.

FLUSH-CRASH-FOUNDER CYCLE [population fluctuation]
= rash-crash-founder cycle

Carson, H. L. 1975. The genetics of speciation at the diploid level. *Amer. Nat.* 109:83–92.

FLYING SQUIRREL THEORY *see* **PARANOTAL THEORY**

FOLD CATASTROPHE [evolution]

Dodson, M. M. 1975. Quantum evolution and the fold catastrophe. *Evol. Theory* 1:107–18.

______ , & A. Hallam. 1977. Allopatric speciation and the fold catastrophe. *Amer. Nat.* 111:415–33.

Cf. THOM'S THEORY OF CATASTROPHES

FOLIAR THEORY [of the ovuliferous scale]

Sachs, J. von. 1868. *Lehrbuch der Botanik*. Leipzig: Engelmann, p. 427. (Transl. 1875 as: *Textbook of Botany*, by A. W. Bennett & W. T. Thiselton-Dyer. Oxford: Clarendon Press.)

Arber, A. 1950. *The Natural Philosophy of Plant Form.* Cambridge: Cambridge University Press, pp. 129*ff.*

Eyde, R. H. 1975. The foliar theory of the flower. *Amer. Sci.* 63:430–37.

FOLIAR THEORY [of the tubular leaf of Nepenthes]

MacFarlane, J. M. 1889. Observations on pitchered insectivorous plants (Part I). *Ann. Bot.* 3:253–66.

Troll, W. 1939. *Vergleichende Morphologie der höheren Pflanzen.* Band 1. *Vegetionsorgane*, teil 2. Berlin: Gebrüder Borntraeger.

Franck, D. H. 1976. The morphological interpretation of epiascidate leaves. *Bot. Rev.* 42:345–88.

FOOD CHAIN [ecology]

Allee, W. C., A. E. Emerson, O. Park, T. Park, & K. P. Schmidt. 1949. *Principles of Animal Ecology.* Philadelphia: Saunders, p. 516.

Phillipson, J. 1966. *Ecological Energetics.* New York: St. Martin's, chap. 2.

FOOD WEB [ecology]

Allee, W. C., A. E. Emerson, O. Park, T. Park, & K. P. Schmidt. 1949. *Principles of Animal Ecology.* Philadelphia: Saunders, p. 516.

Paine, R. T. 1966. Food web complexity and species diversity. *Amer. Nat.* 100:65–75.

Phillipson, J. 1966. *Ecological Energetics.* New York: St. Martin's, chap. 2.

Cohen, J. E. 1978. *Food Webs and Niche Space.* Princeton, N.J.: Princeton University Press.

FORAGING OPTIMIZATION *see* **OPTIMAL FORAGING THEORY**

FORBIDDEN CLONE THEORY or HYPOTHESIS [of immune response]

Burnet, F. M. 1959. Auto-immune disease. II. Pathology of the immune response. *Brit. Med. J.* 2:720–25.

Abramoff, P., & M. F. LaVie. 1970. *Biology of the Immune Response.* New York: McGraw-Hill, p. 247.

FORMATION [ecology]

Clements, F. E. 1928. *Plant Succession and Indicators.* New York: H. W. Wilson, pp. 117*ff.*

FOUNDER PRINCIPLE or EFFECT [evolution]

Mayr, E. 1942. *Systematics and the Origin of Species*. New York: Columbia University Press, p. 237.

______ . 1954. Change of genetic environment and evolution. *Evolution as a Process*. Edited by J. Huxley, A. C. Hardy, & E. B. Ford. London: Allen & Unwin, pp. 157–80.

______ . 1963. *Animal Species and Evolution*. Cambridge: Belknap Press of Harvard University Press, chaps. 8, 17.

Carson, H. L. 1971. Speciation and the founder principle. *Stadler Genet. Symp*. 3:51–70.

Cf. GENETIC DRIFT

FOUNDER-FLUSH SPECIATION THEORY [evolution]

Carson, H. L. 1968. The population flush and its genetic consequences. *Population Biology and Evolution*. Edited by R. C. Lewontin. Syracuse, N.Y.: Syracuse University Press, pp. 123–37.

______ . 1975. Genetics of speciation. *Amer. Nat*. 109:83–92.

Powell, J. R. 1978. The founder-flush speciation theory: an experimental approach. *Evolution* 32:465–74.

Cf. FOUNDER PRINCIPLE

FRASER-DARLING LAW *see* **DARLING EFFECT**

FREE SPACE THEORY *see* **APPARENT FREE SPACE**

FRENCH FLAG PROBLEM [of development]

Wolpert, L. 1968. The French flag problem: a contribution to the discussion on pattern development and regulation. *Towards a Theoretical Biology*. Edited by C. H. Waddington. Edinburgh: Edinburgh University Press, I: 125–33.

______ . 1971. Positional information and pattern formation. *Curr. Top. Dev. Biol*. 6:183–224.

Gould, S. J. 1977. *Ontogeny and Phylogeny*. Cambridge: Harvard University Press, p. 237.

Cf. SPIEGELMAN'S RULE

FREQUENCY, LAW OF [biogeography]

= FREQUENCY-DISTRIBUTION CURVES = Raunkiaer's law of frequency

Raunkiaer, C. 1909–10. Formationsundersögelse og Formations-statistik. *Bot. Tidsskr*. 30:20–132.

______ . 1918. Recherches statistiques sur les formations végétales. *Biol. Meddr*. 1:3.

Kenoyer, L. A. 1927. A study of Raunkiaer's law of frequency. *Ecology* 8:341–49.

Gleason, H. A. 1929. The significance of Raunkiaer's law of frequency. *Ecology* 10:406–08.

McIntosh, R. P. 1962. Raunkiaer's "law of frequency." *Evolution* 43:533–35.

FREQUENCY-DEPENDENT SELECTION [evolution]

Petit, C. 1951. Le rôle de l'isolement sexuel dans l'évolution des populations de *Drosophila melanogaster*. *Bull. Biol. Fr. Belg.* 85:392–418.

Williamson, H. H. 1958. Selection, controlling factors, and polymorphism. *Amer. Nat.* 92:329–35.

Clarke, B., & P. O'Donald. 1964. Frequency dependent selection. *Heredity* 19:201–06.

Kojima, K. 1971. Is there a constant fitness value for a given genotype? No! *Evolution* 25:281–85.

Ayala, F. J., & C. A. Campbell. 1974. Frequency-dependent selection. *Annu. Rev. Ecol. Syst.* 5:115–38.

Gromko, M. H. 1977. What is frequency-dependent selection? *Evolution* 31:438–42.

DeBenedictus, P. A. 1978. Are populations characterized by their genes or by their genotypes? *Amer. Nat.* 112:155–75.

Cf. DENSITY-DEPENDENT SELECTION; DISRUPTIVE SELECTION

FREQUENCY-DISTRIBUTION CURVES [biogeography]

= FREQUENCY, LAW OF = Raunkiaer's law of frequency

Jaccard, P. 1902. Lois de distribution florale dans la zone alpine. *Bull. Soc. Vaud. Sci. Nat.* 38:69–130.

Goodall, D. W. 1952. Quantitative aspects of plant distribution. *Biol. Rev.* 27:194–245.

FROZEN ACCIDENT THEORY [origin of the amino acid code]

Crick, F. H. C. 1968. The origin of the genetic code. *J. Mol. Biol.* 38:367–79.

Jukes, T. H. 1978. The amino acid code. *Adv. Enzymol.* 47:375–432.

FUHRMAN'S RULE [parasitology]

Dogiel, V. A. 1964. *General Parasitology*. Edinburgh: Oliver & Boyd.

Price, P. W. 1977. General concepts on the evolutionary biology of parasites. *Evolution* 31:405–20.

FUNCTIONAL DEMAND THEORY [of growth regulation]

Morgan, T. H. 1901. *Regeneration*. New York: Macmillan.

Swann, M. M. 1958. The control of cell division: a review. II.

Special mechanisms. *Cancer Res.* 18:1118–60.

Goss, R. J. 1964. *Adaptive Growth.* London: Academic Press, chaps. 3, 15.

FUNDAMENTAL LAW OF BIOGENESIS *see* **BIOGENESIS**

FUSION THEORY [of heredity]

Hertwig, O. 1890. Vergleich der Ei- und Samenbildung bei Nematoden. Eine Grundlage für Celluläre Streitfragen. *Arch. Mikrosk. Anat.* 36:1–138.

Churchill, F. B. 1970. Hertwig, Weismann, and the meaning of reduction division circa 1890. *Isis* 61:429–49.

[G]

GALEN'S THEORY [of respiration]

Fleming, D. 1955. Galen on the motions of the blood in the heart and lungs. *Isis* 46:14–21.

Singer, C. 1956. *Galen on Anatomical Procedures.* Oxford: Oxford University Press.

GALTON'S LAW *see* **ANCESTRAL INHERITANCE, LAW OF**

GAMES, THEORY OF [evolution]

Lewontin, R. C. 1961. Evolution and the theory of games. *J. Theor. Biol.* 1:382–403.

______. 1976. Evolution and the theory of games. *Bost. Stud. Philos. Sci.* 27:286–311.

Slobodkin, L. B. 1964. The strategy of evolution. *Amer. Sci.* 52:342–57.

GAMETIC DISEQUILIBRIUM [genetics; evolution]

= linkage disequilibrium

Lewontin, R. C., & K. Kojima. 1960. The evolutionary dynamics of complex polymorphisms. *Evolution* 14:450–72.

Smouse, P. E., & J. V. Neel. 1977. Multivariate analysis of gametic disequilibrium on the Yanomama. *Genetics* 85:733–52.

Hedrick, P., S. Jain, & L. Holden. 1978. Multilocus systems in evolution. *Evol. Biol.* 11:101–84.

GASTREA THEORY [evolution]

Haeckel, E. H. 1874–77. Die Gastraea-Theorie, die phylogenetische classification des Thierreichs und die Homologie der Keimblätter. *Jena. Z. Naturw.* 8:1–55; 9:402–508; 11:1–98. *Arch. Zool. Exper.* 3:239–56. *Q. J. Microsc. Sci.* 14:142–65, 223–47.

Radl, E. 1930. *The History of Biological Theories.* Transl. by E. J. Hatfield. London: Oxford University Press, chap. 12.

Hyman, L. H. 1940. *The Invertebrates.* New York: McGraw-Hill, vol. I, pp. 250*ff*.

Dodson, E. O., & P. Dodson. 1976. *Evolution: Process and Product.* New York: Van Nostrand, pp. 164–66.

Cf. BILATEROGASTREA THEORY

GATES OF ANGIOSPERMY [biogeography]

Croizat, L. 1952. *Manual of Phytogeography.* The Hague: Junk.

Ball, I. R. 1975. Nature and formulation of biogeographical hypotheses. *Syst. Zool.* 24:407–30.

Cf. CENTERS OF ORIGIN; GENERALIZED TRACKS

GAUSE'S PRINCIPLE or HYPOTHESIS or AXIOM *see* **COMPETITIVE EXCLUSION**

GAUSE-VOLTERRA, PRINCIPLE OF *see* **COMPETITIVE EXCLUSION**

GEMINATE SPECIES, LAW OF *see* **JORDAN'S LAW**

GENE [genetics]

Morgan, T. H. 1917. The theory of the gene. *Amer. Nat.* 51:513–44.

———. 1926. *The Theory of the Gene.* New Haven, Conn.: Yale University Press.

East, E. M. 1929. The concept of the gene. *Proceedings of the International Congress of Plant Sciences,* Ithaca, N.Y. Aug. 16–23, 1926. Edited by B. M. Duggar. 2 vols. Menasha, Wis.: Banta, 1:889–95.

Goldschmidt, R. B. 1938. *Physiological Genetics.* New York: McGraw-Hill, chap. 4.

Muller, H. J. 1951. The development of the gene theory. *Genetics in the 20th Century.* Edited by L. C. Dunn. New York: Macmillan, pp. 77–99.

Carlson, E. A. 1966. *The Gene: A Critical History*. Philadelphia: Saunders.

______ . 1971. An unacknowledged founding of molecular biology: H. J. Muller's contributions to gene theory, 1910–1936. *J. Hist. Biol.* 4:149–70.

Gilbert, S. F. 1978. The embryological origins of the gene theory. *J. Hist. Biol.* 11:307–51.

GENE CENTERS [genetics; evolution]

= centers of origin

Vavilov, N. I. 1926. *Studies on the Origin of Cultivated Plants*. Leningrad.

Stebbins, G. L. 1950. *Variation and Evolution in Plants*. New York: Columbia University Press, p. 532.

Harlan, J. R. 1951. Anatomy of gene centers. *Amer. Nat.* 85:79–103.

Leppik, E. E. 1970. Gene centers of plants as sources of disease resistance. *Annu. Rev. Phytopath.* 8:323–44.

Cf. AGE AND AREA; MATTHEW'S HYPOTHESIS

GENE CONVERSION GUILDS [genetics]

Atsatt, P. R., & D. J. O'Dowd. 1976. Plant defense guilds. *Science* 193:24–29.

GENE CONVERSION [in intragenic recombination]

= nonreciprocal combination = Mitchell effect = transmutation = transreplication = intragenic recombination = defective copying = reversion = heteroallelic regeneration = asymmetrical crossing-over

Winkler, H. 1930. *Die Konversion der Gene*. Jena: Fischer.

Stern, C. 1932. Über die konversionstheorie. *Biol. Zbl.* 52:367–79.

Lindegren, C. C. 1953. Gene conversion in *Saccharomyces*. *J. Genet.* 51:625–37.

Leblon, G. 1972. Mechanisms of gene conversion in *Abscobolus immersus*. I & II. *Mol. Gen. Genet.* 115:36–48; 116:322–35.

Stadler, D. R. 1973. The mechanism of intragenic recombination. *Annu. Rev. Genet.* 7:113–27.

Kushev, V. V. 1974. *Mechanisms of Genetic Recombination*. New York: Consultants Bureau, chap. 3.

Broker, T. R., & A. H. Doermann. 1975. Molecular and genetic recombination of bacteriophage T4. *Annu. Rev. Genet.* 9:213–44.

Hastings, P. J. 1975. Some aspects of recombination in eukaryotic organisms. *Annu. Rev. Genet.* 9:129–44.

GENE DUPLICATION [evolution]

Ohno, S. 1970. *Evolution by Gene Duplication*. New York: Springer-Verlag.

McLachlan, A. D. 1972. Repeating sequences and gene duplication in proteins. Appendix-substitution frequencies in proteins. *J. Mol. Biol.* 64:417-33, 434–38.

Frazzetta, T. H. 1975. *Complex Adaptations in Evolving Populations*. Sunderland, Mass.: Sinauer Associates, chap. 4.

GENE FLOW [population genetics]

Birdsell, J. B. 1950. Some implications of the genetical concept of race in terms of spatial analysis. *Cold Spr. Harb. Symp. Quant. Biol.* 15:259–314.

Mayr, E. 1963. *Animal Species and Evolution*. Cambridge: Belknap Press of Harvard University Press.

Erlich, P. R., & P. H. Raven. 1969. Differentiation of populations. *Science* 165:1228–32.

Endler, J. A. 1973. Gene flow and population differentiation. *Science* 179:243–50.

GENE FLOW-VARIATION HYPOTHESIS [of population diversity]

Soule, M. 1971. The variation problem: the gene flow-variation hypothesis. *Taxon* 20:37–50.

Zengerl, A. R., S. T. A. Pickett, & F. A. Bayzay. 1977. Some hypotheses on variation in plant populations and an experimental approach. *Biologist* 59:113–22.

GENE-FOR-GENE CONCEPT [host-parasite relationship]

Flor, H. H. 1955. Host-parasite interaction in flax rust—its genetic and other implications. *Phytopathology* 45:680–85.

———. 1956. The complementary genetic system in flax rust. *Adv. Genet.* 8:29–59.

———. 1971. Current status of the gene-for-gene concept. *Annu. Rev. Phytopath.* 9:275–96.

Person, C. 1959. Gene-for-gene relationships in host: parasite systems. *Can. J. Bot.* 37:1101–30.

———. 1962. The gene-for-gene concept. *Nature* 194:561–62.

Cf. COEVOLUTION

GENE POOL [population genetics; evolution]

Dobzhansky, 1950. Mendelian populations and their evolution. *Amer. Nat.* 84:401–18.

———. 1951. *Genetics and the Origin of Species*. 3d ed., rev.

New York: Columbia University Press, pp. 155*ff*.

Erlich, P. R., & R. W. Holm. 1963. *The Process of Evolution*. New York: McGraw-Hill.

Mayr, E. 1963. *Animal Species and Evolution*. Cambridge: Belknap Press of Harvard University Press.

Harlan, J. R., & J. M. J. deWet. 1971. Toward a rational classification of cultivated plants. *Taxon* 20:509–17.

GENE REGULATION THEORY [of gene activity and differentiation]

Zuckerkandl, E. 1964. Controller-gene diseases. *J. Mol. Biol.* 8:128–47.

Britten, R. J., & E. H. Davidson. 1969. Gene regulation for higher cells: a theory. *Science* 165:349–57.

______. 1971. Repetitive and non-repetitive DNA sequences and a speculation on the origins of evolutionary novelty. *Q. Rev. Biol.* 46:11–38.

Englesberg, E., & G. Wilcox. 1974. Regulation: positive control. *Annu. Rev. Genet.* 8:219–42.

Düygünes, N. 1975. On the theories of gene regulation and differentiation in eukaryotes. *Acta Biotheor.* 24:120–26.

Cf. OPERON THEORY; REGULATOR GENES

GENECOLOGY

Turesson, G. 1923. The scope and impact of genecology. *Hereditas* 4:171–76.

Heslop-Harrison, J. 1964. Forty years of genecology. *Adv. Ecol. Res.* 2:159–247.

Langlet, O. 1971. Two hundred years genecology. *Taxon* 20:653–722.

GENERALIZED TRACK [biogeography]

Croizat, L., G.J. Nelson, & D.E. Rosen. 1974. Centers of origin and related concepts. *Syst. Zool.* 23:265–87.

McDowall, R.M. 1978. Generalized tracks and dispersal in biogeography. *Syst. Zool.* 27:88–104.

Craw, R.C. 1979. Generalized tracks and dispersal in biogeography: a response to R.M. McDowall. *Syst. Zool.* 28:99–107.

GENERATION LAW [evolution]

Thompson, W. R. 1931. On the reproduction of organisms with overlapping generations. *Bull. Entomol. Res.* 22:147–72.

Cole, L. C. 1954. The population consequences of life history phenomena. *Q. Rev. Biol.* 29:103–37.

GENERIC CYCLES, THEORY OF [biogeography]

Good, R. D'O. 1930. The geography of the genus *Coriaria*. *New Phytol*. 29:170–98.

Cain, S. A. 1944. *Foundations of Plant Geography*. New York: Harper, pp. 240–41.

Good, R. 1974. *The Geography of the Flowering Plants*. 4th ed. London: Longman, pp. 44–45.

GENES-FIRST HYPOTHESIS [origin of life]

Lehninger, A. L. 1970. *Biochemistry*. New York: Worth, p. 782.

Moody, A. R. 1970. *Introduction to Evolution*. 3d ed. New York: Harper & Row.

GENETIC ASSIMILATION [evolution]

= threshold selection

Waddington, C. H. 1953. The "Baldwin effect," "genetic assimilation" and "homeostasis." *Evolution* 7:386–87.

——— . 1959. Evolutionary systems: animal and human. *Nature* 183:1634–38.

——— . 1961. Genetic assimilation. *Adv. Genet.* 10:257–93.

Stern, C. 1958. Selection for subthreshold differences and the origin of pseoduexogenous adaptations. *Amer. Nat.* 92:313–16.

Cf. BALDWIN EFFECT

GENETIC BALANCE *see* **GENIC-BALANCE HYPOTHESIS**

GENETIC CONTROL HYPOTHESIS [of population cycles]

Chitty, D. 1960. Population processes in the vole and their relevance to general theory. *Can. J. Zool.* 38:99–113.

——— . 1967. The natural selection of self-regulatory behavior in animal populations. *Proc. Ecol. Soc. Austr.* 2:51–78.

Pianka, E. R. 1978. *Evolutionary Ecology*. 2d ed. New York: Harper & Row, p. 125.

GENETIC DEATH

Muller, H. J. 1950. Our load of mutations. *Amer. J. Hum. Genet.* 2:111–76.

Cf. GENETIC LOAD; HALDANE'S DILEMMA

GENETIC DISTANCE

Wright, S. 1943. Isolation by distance. *Genetics* 31:114–38.

Nei, M. 1972. Genetic distance between populations. *Amer. Nat.* 106:283–92.

______. 1975. *Molecular Population Genetics and Evolution.* New York: American Elsevier, pp. 175*ff.*

Cf. ISOLATION BY DISTANCE

GENETIC DRIFT

= drift = Gulick effect

Wright, S. 1921. Systems of mating. I. The biometric relations between parent and offspring. *Genetics* 6:111–23.

______. 1931. Evolution in Mendelian populations. *Genetics* 16:97–159.

______. 1970. Random drift and the shifting blance theory of evolution. *Mathematical Topics in Population Genetics.* Edited by K. Kojima. Berlin: Springer-Verlag, pp. 1–31.

Dobzhansky, T. 1951. *Genetics and the Origin of Species.* 3d ed. New York: Columbia University Press, pp. 156*ff.*

Mayr, E. 1963. *Animal Species and Evolution.* Cambridge: Belknap Press of Harvard University Press, pp. 204*ff.*

Lande, R. 1976. Natural selection and random genetic drift in phenotypic evolution. *Evolution* 30:314–34.

Cf. BALANCE-SHIFT THEORY; SEWALL WRIGHT EFFECT

GENETIC EQUILIBRIUM *see* **HARDY-WEINBERG LAW**

GENETIC FEEDBACK [population genetics]

Pimentel, D. 1961. Animal population regulation by the genetic feed-back mechanism. *Amer. Nat.* 95:65–79.

______. 1968. Population regulation and genetic feedback. *Science* 159:1432–37.

Levin, S. A. 1972. A mathematical analysis of the genetic feedback mechanism. *Amer. Nat.* 106:145–64.

Cf. FEEDBACK CONTROL

GENETIC HOMEOSTASIS

= genetic inertia

Darlington, C. D., & K. Mather. 1949. *The Elements of Genetics.* New York: Macmillan, pp. 296*ff.*

Lerner, I. M. 1954. *Genetic Homeostasis.* New York: Wiley.

Lewontin, R. C. 1956. Studies on homeostasis and heterozygosity. I. General considerations. *Amer. Nat.* 90:237–55.

GENETIC HOMOGENEITY [evolution]

Stebbins, G. L. 1942. The genetic approach to problems of rare and endemic species. *Madroño* 6:241–58.

GENETIC LOAD

Muller, H. J. 1950. Our load of mutations. *Amer. J. Hum. Genet.* 2:111–76.

Crow, J. F. 1958. Some possibilities for measuring selection intensities in man. *Hum. Biol.* 30:1–13.

———. 1970. Genetic loads and the cost of natural selection. *Mathematical Topics in Population Genetics.* Edited by K. Kojima. Berlin: Springer-Verlag, pp. 128–77.

———. & M. Kimura. 1965. The theory of genetic loads. *Proc. 11th Int. Congr. Genet.*, The Hague, Netherlands 3:495–505.

Kimura, M., & J. F. Crow. 1964. The number of alleles that can be maintained in a finite population. *Genetics* 49:725–38.

Wallace, B. 1970. *Genetic Load: Its Biological and Conceptual Aspects.* Englewood Cliffs, N.J.: Prentice-Hall.

Taylor, C. E. 1975. Genetic loads in heterogeneous environments. *Genetics* 80:621–35.

GENETIC MODES, THEORY OF [evolution]

Baldwin, J. M. 1902. *Development and Evolution.* New York: Macmillan, p. 300.

Cf. EMERGENT EVOLUTION

GENETIC POLYMORPHISM, THEORY OF

Ford, E. B. 1940. Polymorphism and taxonomy. *The New Systematics.* Edited by J. Huxley. Oxford: Clarendon Press, pp. 493–513.

———. 1965. *Genetic Polymorphism.* Cambridge: M. I. T. Press.

———. 1971. *Ecological Genetics.* 3d ed. London: Chapman & Hall, chap. 6.

Muller, H. J. 1950. Our load of mutations. *Amer. J. Hum. Genet.* 2:111–76.

———. 1959. The mutation theory re-examined. *Proc. 10th Int. Congr. Genet.*, Montreal 1:306–17.

Huxley, J. S. 1955. Morphism and evolution. *Heredity* 9:1–52.

GENETIC REVOLUTIONS [evolution]

Mayr, E. 1954. Change of genetic environment and evolution. *Evolution as a Process.* Edited by J. Huxley, A. C. Hardy & E. B. Ford. London: Allen & Unwin, pp. 157–80.

———. 1963. *Animal Species and Evolution.* Cambridge: Belknap Press of Harvard University Press, chap. 17.

GENETIC "THEORY OF RELATIVITY" [population genetics]

Mayr, E. 1955. Integration of genotypes: synthesis. *Cold Spr. Harb. Symp. Quant. Biol.* 20:327–33.

GENETIC VARIABILITY

Dobzhansky, T., H. Levene, B. Spassky, & N. Spassky. 1959. Release of genetic variability through recombination. III. *Drosophila prosaltans. Genetics* 44:75-92.

Simmonds, N.W. 1962. Variability in crop plants, its use and conservation. *Biol. Rev.* 37:442–65.

Mayr, E. 1963. *Animal Species and Evolution.* Cambridge: Belknap Press of Harvard University Press, chaps. 7–9.

Smithies, O. 1968. Genetics of variability. *Brookh. Symp. Biol.* 21:243–58.

Bernstein, S. C., L. H. Throckmorton, & J. L. Hubby. 1973. Still more genetic variability in natural populations. *Proc. Natl. Acad. Sci.* 70:3928–31.

Selander, R. K., & W. E. Johnson. 1973. Genetic variations among vertebrate species. *Annu. Rev. Ecol. Syst.* 4:75–91.

Bowler, P. J. 1974. Darwin's concept of variation. *J. Hist. Med.* 29:196–212.

Lewontin, R. C. 1974. *The Genetic Basis of Evolutionary Change.* New York: Columbia University Press.

GENETICAL RECOMBINATION

Owen, A. R. G. 1950. The theory of genetical recombination. *Adv. Genet.* 3:117–57.

Cf. GENE CONVERSION

GENETICAL THEORY [evolution of altruistic behavior]

Hamilton, W. D. 1964. The genetical evolution of social behaviour. I. *J. Theor. Biol.* 7:1–16.

Eberhard, M. J. W. 1975. The evolution of social behavior by kin selection. *Q. Rev. Biol.* 50:1–33.

Cf. KIN SELECTION

GENIC-BALANCE HYPOTHESIS [sex-determination in haplodiploidy]

= genetic balance

Bridges, C. B. 1922. The origin of variations in sexual and sex-linked characters. *Amer. Nat.* 56:51–63.

Cunha, A. B. da, & W. E. Kerr. 1957. A genetical theory to explain sex-determination by arrhenotokous parthenogenesis. *Forma Functio* 1(4):33–36.

Crozier, R. H. 1971. Heterozygosity and sex determination in haplo-diploidy. *Amer. Nat.* 105:399–412.

Cf. MULTIPLE-ALLELE, SINGLE-LOCUS HYPOTHESIS; MULTIPLE-LOCUS HYPOTHESIS

GENOME SYSTEMATIZATION THEORY [evolution]

Riedl, R. 1977. A systems-analytical approach to macro-evolutionary phenomena. *Q. Rev. Biol.* 52:351–70.

GENORHEITRON [evolution]

= genorheithrum

Lam, H. J. 1938. Studies in phylogeny. I. On the relation of taxonomy, phylogeny and biogeography. II. On the phylogeny of the Malaysian Burseraceae-Canarieae in general and *Haplolobus* in particular. *Blumea* 3:114–58.

Croizat, L. 1952. *Manual of Phytogeography*. The Hague: Junk, chap. 2.

Good, R. 1974. *The Geography of the Flowering Plants*. 4th ed. London: Longman, pp. 42–43.

GENOTYPE [genetics]

Johannsen, W. 1909. *Elemente der exakten Erblichkeitslehre*. Jena: Fischer.

——— . 1911. The genotype conception of heredity. *Amer. Nat.* 45:129–59.

East, E. M. 1911. The genotype hypothesis and hybridization. *Amer. Nat.* 45:160–74.

Turesson, G. 1922. The species and the variety as ecological units. *Hereditas* 3:100–13.

Allen, C. E. 1942. Regeneration, development and genotype. *Amer. Nat.* 76:225–38.

Churchill, F. B. 1974. William Johannsen and the genotype concept. *J. Hist. Biol.* 7:5–30.

Wanscher, J. H. 1975. The history of Wilhelm Johannsen's genetical terms and concepts from the period 1903 to 1926. *Centaurus* 19:125–47.

GENUS, CONCEPT OF [systematics]

Bartlett, H. H. 1940. The history of the generic concept in botany. *Bull. Torr. Bot. Club* 67:349–62.

Clayton, W. D. 1972. Some aspects of the genus concept. *Kew Bull.* 27:281–87.

GEOGRAPHIC CHARACTER GRADIENT *see* **CLINE**

GEOGRAPHIC ISOLATION [evolution]

Mayr, E. 1959. Isolation as an evolutionary factor. *Proc. Amer. Philos. Soc.* 103:221–30.

Lesch, J. E. 1975. The role of isolation in evolution: George J. Romanes and John T. Gulick. *Isis* 66:483–503.

Cf. REPRODUCTIVE ISOLATION

GEOGRAPHIC SPECIATION[evolution]

Wagner, M. 1889. *Die Enstehung der Arten durch räumliche Sonderung. Gesammelte Aufsätze.* Basel: Schwabe.

Mayr, E. 1959. Isolation as an evolutionary factor. *Proc. Amer. Philos. Soc.* 103:221–30.

______ . 1963. *Animal Species and Evolution.* Cambridge: Belknap Press of Harvard University Press, chap. 16.

Endler, J. A. 1977. *Geographic Variation, Speciation, and Clines.* Princeton, N.J.: Princeton University Press.

Cf. ALLOPATRIC SPECIATION; GEOGRAPHIC ISOLATION; JORDAN'S LAW

GEOGRAPHICAL PARTHENOGENESIS [evolution]

Vandel, A. 1931. *La Parthénogenèse.* Paris: Doin Freres.

Suomalainen, E. 1962. Significance of parthenogenesis in the evolution of insects. *Annu. Rev. Entomol.* 7:349–66.

White, M. J. D. 1973. *Animal Cytology and Evolution.* 3d ed. Cambridge: Cambridge University Press, chap. 19.

Parker, E. D., Jr., R. K. Selander, R. O. Hudson, & L. J. Lester. Genetic diversity in colonizing parthenogenetic cockroaches. *Evolution* 31:836–42.

Cf. THELYTOKY

GEOLOGICAL TIME THEORY [of insect-plant coevolution]

Southwood, T. R. E. 1961. The numbers of species of insects associated with various trees. *J. Anim. Ecol.* 30:1–8.

Opler, P. A. 1974. Oaks as evolutionary islands for leaf-mining insects. *Amer. Sci.* 62:67–73.

Claridge, M. F., & M. R. Wilson. 1978 . British insects and trees: a study in island biogeography or insect/plant coevolution. *Amer. Nat.* 112:451–56.

GEOPHYTOPATHOLOGY

= geoplantpathology = pathogeography

Reichart, I. 1953. A biogeographical approach to phytopathology. *Proc. 12th Int. Bot. Congr.*, Stockholm, 1950:730–31.

Waltzien, H. C. 1967. Geopathologie der Pflanzen. *Z. Pflanzenkr.* 74:176–89.

———. 1972. Geophytopathology. *Annu. Rev. Phytopath.* 10:277–98.

GEOPLANTPATHOLOGY *see* **GEOPHYTOPATHOLOGY**

GERM THEORY OF DISEASE

Prévost, P. 1820. *Notice de la vie et des écrits d'Isaac-Bénédict Prévost.* Geneva: Paschound.

DeBary, A. 1853. *Untersuchungen über die Brandpilze und die durch sie verursachten Krankheiten der Pflanzen mit Rücksicht auf das Getreide und andere Nutzpflanzen.* Berlin: Müller.

Buller, A. H. R. 1915. Micheli and the discovery of reproduction in fungi. *Trans. Roy. Soc. Can.* IV 9:1–25.

Keitt, G. W. 1959. History of plant pathology. *Plant Pathology, an Advanced Treatise.* Edited by J. G. Horsfall & A. E. Dimond. vol. I: *The Diseased Plant.* New York: Academic Press, pp. 61–97.

Cf. DEFICIENCY DISEASE CONCEPT

GERM-LAYER THEORY [morphogenesis]

Baer, K. E. von. 1828–37. *Uber Entwickelungsgeschichte der Thiere.* Königsberg: Gebrüder Borntraeger.

de Beer, G. R. 1951. *Embryos and Ancestors.* Oxford: Clarendon Press, p. 127.

Oppenheimer, J. 1940. The non-specificity of the germ layers. *Q. Rev. Biol.* 15:1–27.

Baxter, A. L. 1977. E. B. Wilson's "Destruction of the Germ-Layer Theory." *Isis* 68:363–74.

GERMPLASM THEORY [of inheritance]

Weismann, A. 1892. *Das Keimplasma. Eine Theorie der Vererbung.* Jena: Fischer. (*The Germ-Plasm. A Theory of Heredity.* Transl. by W. N. Parker & H. Rönnefeldt. 1893. London: Walter Scott.)

Berrill, N.J., & C. K. Liu. 1948. Germplasm, Weismann, and hydrozoa. *Q. Rev. Biol.* 23:124–32.

Churchill, F. B. 1968. August Weismann and a break from tradition. *J. Hist. Biol.* 1:91–112.

GILL THEORY [of insect wing origin]

Snodgrass, R. E. 1958. Evolution of arthropod mechanisms. *Smithson. Misc. Coll.* 138:1–77.

Wigglesworth, V. B. 1976. The evolution of insect flight. *Insect Flight.* Edited by R. C. Rainey. Oxford: Blackwell, pp.

255–69. (*Symp. Roy. Entomol. Soc. Lond.* no. 7.)

Kukalova-Peck, J. 1978. Origin and evolution of insect wings and their relation to metamorphosis, as documented by the fossil record. *J. Morph.* 156:53–125.

Daly, H. V., J. T. Doyen, & P. R. Ehrlich. 1978. *Introduction to Insect Biology and Diversity*. New York: McGraw-Hill, p. 267.

Cf. PARANOTAL THEORY

GILLESPIE-LANGLEY HYPOTHESIS or MODEL [enzyme variation in natural populations]

Gillespie, J. 1974. A general model to account for genetic variation in natural populations. *Genetics* 76:837–48.

______ . & C. H. Langley. 1974. The role of environmental grain in the maintenance of genetic variation. *Amer. Nat.* 108:831–36.

GLACIAL CONTROL THEORY [of coral reef formation]

Daly, R. A. 1910. Pleistocene glaciation and the coral reef problem. *Amer. J. Sci.* ser. 4, 30:297–308.

Stoddart, D. R. 1969. Ecology and morphology of recent coral reefs. *Biol. Rev.* 44:433–98.

GLEASON'S INDIVIDUALISTIC CONCEPT *see* **INDIVIDUALISTIC CONCEPT**

GLOGER'S RULE [of melanin pigmentation]

Gloger, C. L. 1833. *Das Abändern der Vögel durch Einfluss des Klimas*. Breslau: Schultz.

Searle, A. G. 1968. *Comparative Genetics of Coat Color in Mammals*. London: Logos.

Rappoport, E. H. 1969. Gloger's rule and pigmentation of collembola. *Evolution* 23:622–26.

GLUTTONY PRINCIPLE [ethology]

Brown, J. L. 1975. *The Evolution of Behavior*. New York: Norton, p. 143.

GOETHE–DE CANDOLLE THEORY [of the gynaeceum]

= classical theory = Goethe's theory of metamorphosis [carpel evolution]

Goethe, J. W. von. 1790. *Versuch die Metamorphose der Pflanzen zu Erklaren*. Gotha: Ettinger. (Transl. by A. Arber. 1946. *Goethe's Botany*. Waltham, Mass.: Chronica Botanica. vol. 10, no. 2, pp. 63–126.)

Candolle, A. P. de. 1827. *Organographie Vegetale.* Tome l. Paris: Deterville. (Transl. by H. Kingdon. 1841. *Vegetable Organography.* London: Houlston & Stoneman.)

Arber, A. 1937. The interpretation of the flower: a study of some aspects of morphological thought. *Biol. Rev.* 12:157–84.

______. 1950. *The Natural Philosophy of Plant Form.* Cambridge: Cambridge University Press, pp. 41*ff.*

Wardlaw, C. W. 1965. *Organization and Evolution in Plants.* London: Longmans, pp. 19*ff.*

Cf. CLASSICAL SHOOT THEORY

GOETHE'S SPIRAL TENDENCY *see* **PHYLLOTAXY**

GOETHE'S VERTEBRAL THEORY [of the origin of the skull]

Goethe, J. W. von. 1820. *Naturwissenschaft Überhaupt, besonders zur Morphologie.* I.

Huxley, T. H. 1858. The Croonian lecture—on the theory of the vertebrate skull. *Proc. Roy. Soc. Lond.* 9:381–457.

Singh-Roy, K. K. 1967. On Goethe's vertebral theory of origin of the skull, a recent approach. *Anat. Anz.* 120:250–59.

GONOPHYLL THEORY [morphology evolution]

Melville, R. 1960. A new theory of the angiosperm flower. *Nature* 188:14–18.

______. 1962. A new theory of the angiosperm flower. I. The gynoecium. *Kew Bull.* 16:1–50.

______. 1963. A new theory of the angiosperm flower. II. The androecium. *Kew Bull.* 17:1–63.

GOODNESS-OF-FIT TEST [population genetics]

Wallace, B. 1958. The comparison of observed and calculated zygotic distributions. *Evolution* 12:113–15.

Lewontin, R. C., & C. C. Cockerham. 1959. The goodness-of-fit test for detecting natural selection in random mating populations. *Evolution* 13:561–64.

Li, C. C. 1959. Notes on relative fitness of genotypes that forms a geometric progression. *Evolution* 13:546–67.

GOSSE'S LAW *see* **PROCHRONISM, LAW OF**

GRADIENT ANALYSIS [ecology]

Ramensky, L. G. 1924. Die Grundgesetzmässigkeiten im Aufbau der Vegetationsdecke. (In Russian.) Wjestn. Opytn.

djela Woronesch. 37 pp. (German summary: *Bot. Zblt.*, N. F. 7:453–55, 1926.)

Whittaker, R. H. 1967. Gradient analysis of vegetation. *Biol. Rev.* 42:207–64.

Cf. CONTINUUM CONCEPT; INDIVIDUALISTIC CONCEPT [of the plant association]

GRADIENT THEORY [morphogenesis]

= Child's gradient theory

Child, C. M. 1941. *Patterns and Problems of Development.* Chicago: University of Chicago Press.

Prat, H. 1948. Histo-physiological gradients and plant organogenesis. *Bot. Rev.* 14:603–43.

Wardlaw, C. W. 1952. *Phylogeny and Morphogenesis.* London: Macmillan, p. 442.

______. 1965. *Organization and Evolution in Plants.* London: Longmans, pp. 27-29.

Cf. FIELD CONCEPT

GRADUALISM *see* **PHYLETIC GRADUALISM**

GRAIN, FINE- AND COARSE- [evolution; ecology]

Levins, R. 1968. *Evolution in Changing Environments.* Princeton, N.J.: Princeton University Press, chap. 4.

Gillespie, J. 1973. Polymorphism in random environments. *Theor. Pop. Biol.* 4:193–95.

______. 1974. The role of environmental grain in the maintenance of genetic variation. *Amer. Nat.* 108:831–36.

GRAPHICAL EXPLOITATION THEORY *see* **BIOLOGICAL EXPLOITATION THEORY**

GRAPHICAL THEORY [of predation]

Rosenzweig, M. L. 1969. Why the prey curve has a hump. *Amer. Nat.* 103:81–87.

______. & R. H. MacArthur. 1963. Graphical representation and stability conditions of predator-prey interactions. *Amer. Nat.* 97:209–23.

GREEN-BENSON HYPOTHESIS [of membrane structure]

Green, D. E., & J. F. Perdue. 1966. Membranes as expressions of repeating units. *Proc. Natl. Acad. Sci.* 55:1295–302.

Benson, A. A. 1966. On the orientation of lipids in chloroplast and cell membranes. *J. Amer. Oil Chem. Soc.* 43:265–70.

Nystrom, R. A. 1973. *Membrane Physiology.* Englewood Cliffs, N.J.: Prentice-Hall, pp. 35–40.

GREGG'S PARADOX [of taxonomic classification]

Gregg, J. R. 1954. *The Language of Taxonomy*. New York: Columbia University Press.

Ruse, M. E. 1971. Gregg's paradox: a proposed revision to Buck and Hull's solution. *Syst. Zool.* 20:239–45.

Løvtrup. S. 1974. *Epigenetics*. New York: Wiley, p. 478.

GRINNEL'S AXIOM *see* **COMPETITIVE EXCLUSION**

GROUP SELECTION [evolution]

Wright, S. 1945. Tempo and mode in evolution: a critical review. *Ecology* 26:415–19.

Dunbar, M. J. 1960. The evolution of stability in marine environments: natural selection at the level of the ecosystem. *Amer. Nat.* 94:129–36.

——— . 1968. *Ecological Development in Polar Regions*. Englewood Cliffs, N.J.: Prentice-Hall.

Wynne-Edwards, V. C. 1962. *Animal Dispersion in Relation to Social Behaviour*. Edinburgh: Oliver & Boyd.

——— . 1964. Group selection and kin selection. *Nature* 201:1147.

Maynard Smith, J. 1964. Group selection and kin selection. *Nature* 201:1145–47.

——— . 1976. Group selection. *Q. Rev. Biol.* 51:277–83.

Williams. G. C. 1966. *Adaptation and Natural Selection: A Critique of Some Current Evolutionary Thought*. Princeton, N.J.: Princeton University Press, chap. 4.

Lewontin, R. C. 1970. The units of selection. *Annu. Rev. Ecol. Syst.* 1:1–18.

——— . 1974. *The Genetic Basis of Evolutionary Change*. New York: Columbia University Press.

Van Valen, L. 1975. Group selection, sex, and fossils. *Evolution* 29:87–94.

Wade, M. J. 1977. An experimental study of group selection. *Evolution* 31:134–53.

——— . 1978. A critical review of the models of group selection. *Q. Rev. Biol.* 53:101–14.

Alexander, R. D., & G. Borgia. 1978. Group selection, altruism, and the levels of organization of life. *Annu. Rev. Ecol. Syst.* 9:449–74.

Cf. KIN SELECTION

GROWTH, LAWS OF

Dickinson, T. A. 1978. Epiphylly in angiosperms. *Bot. Rev.* 44:181–232.

 GROWTH FORCE *see* **MATERIAL COMPENSATION, PRINCIPLE OF**

GROWTH LINE FORMATION, THEORY OF [in mollusc shells]

Lutz, R. A., & D. C. Rhoads. 1977. Anaerobiosis and a theory of growth line formation. *Science* 198:1222–26.

GUILD DEFENSE [ecology]

Atsatt, P. R., & D. J. O'Dowd. 1976. Plant defense guilds. *Science* 193:24–29.

O'Dowd, D.J., & G.B. Williamson. 1979. Stability conditions in plant defense guilds. *Amer. Nat.* 114:379–83.

GULICK EFFECT [evolution]

Gulick, J. T. 1888. Divergent evolution through cumulative segregation. *J. Linn. Soc. Lond. Zool.* 20:189–274.

Kellogg, V. L. 1907. *Darwinism To-Day*. New York: Holt, pp. 249*ff*.

Cf. GENETIC DRIFT

GURWITCH'S THEORY *see* **FIELD CONCEPT**

[H]

HABITAT [ecology]

Yapp, R. H. 1922. The concept of habitat. *J. Ecol.* 10:1–17.

Davis, J. H. 1960. Proposals concerning the concept of habitat and a classification of types. *Ecology* 41:537–41.

Whittaker, R. H., S. A. Levin, & R. B. Root, 1973. Niche, habitat, and ecotope. *Amer. Nat.* 107:321–38.

Doyle, R. W. 1976. Analysis of habitat loyalty and habitat preference in the settlement behavior of planktonic marine larvae. *Amer. Nat.* 110:719–30.

HABITAT-DIVERSITY HYPOTHESIS [species numbers/area]

Williams, C. B. 1964. *Patterns in the Balance of Nature*. London: Academic Press, chap. 7.

Conner, E. F., & E. D. McCoy. 1979. The statistics and biology of the species-area relationship. *Amer. Nat.* 113:791–833.

HABITAT LOYALTY [ecology]

Doyle, R. W. 1976. Analysis of habitat loyalty and habitat preference in the settlement behavior of planktonic marine larvae. *Amer. Nat.* 110:719–30.

HABITAT SELECTION [ecology]

Lack, D. 1940. Habitat selection and speciation in birds. *Brit. Birds* 34:80–84.

Maynard Smith, J. 1966. Sympatric speciation. *Amer. Nat.* 100:637–50.

Meadows, P. S., & J. I. Campbell. 1972. Habitat selection and animal distribution in the sea: the evolution of a concept. *Proc. Roy. Soc. Edinb. B* 73:145–57.

HABITAT SHIFT [ecology]

Schoener, T. W. 1975. Presence and absence of habitat shift in some widespread lizard species. *Ecol. Monogr.* 45:233–58.

Cf. NICHE SHIFT

HAECKEL'S LAWS OF HEREDITY *see* **BIOGENESIS**

HAGEDOORN EFFECT [random survival of variations in evolution]

= reduction of polymorphy, law of

Hagedoorn, A. L., & A. C. Hagedoorn. 1921. *The Relative Value of the Processes Causing Evolution.* The Hague: Martinus Nijhoff.

Fisher, R. A. 1922. On the dominance ratio. *Proc. Roy. Soc. Edinb.* 42:321–41.

Cain, S. A. 1944. *Foundations of Plant Geography.* New York: Harper, p. 295.

HALDANE'S DILEMMA [evolutionary rates]

Haldane, J. B. S. 1957. The cost of natural selection. *J. Genet.* 55:511–24.

Van Valen, L. 1963. Haldane's dilemma, evolutionary rates, and heterosis. *Amer. Nat.* 97:185–90.

Maynard Smith, J. 1968. "Haldane's dilemma" and the rate of evolution. *Nature* 219:1114–16.

O'Donald, P. 1969. "Haldane's dilemma" and the rate of natural selection. *Nature* 221:815–16.

Moran, P. A. P. 1970. "Haldane's dilemma" and the rate of evolution. *Ann. Hum. Genet.* 33:245–49.

HALDANE'S LAW or RULE [of hybrid weakness]

Haldane, J. B. S. 1922. Sex-ratio and unisexual sterility in hybrid animals. *J. Genet.* 12:101–09.

Muller, H. J. 1940. Bearings of the *Drosophila* work on systematics. *The New Systematics.* Edited by J. S. Huxley. Oxford: Clarendon Press, pp. 185–268.

Stebbins, G. L. 1958. The inviability, sterility and weakness of interspecific hybrids. *Adv. Genet.* 9:147–215.

HALDANE'S PRINCIPLE [of mean selective value]

Haldane, J. B. S. 1937. The effect of variation on fitness. *Amer. Nat.* 71:337–49.

HALLER'S RULE [of relative growth]

Haller, A. von. 1762. *Elementa physiologiae corporis humanis.* tom. (IV. Lausanne: Sumptibus Bousquet.)

Rensch, B. 1956. Increase of learning capability with increase of brain size. *Amer. Nat.* 90:81–95.

HANDICAP PRINCIPLE [of sexual selection]

Zahavi, A. 1975. Mate selection—a selection for a handicap. *J. Theor. Biol.* 53:205–14.

Maynard Smith, J. 1976. Sexual selection and the handicap principle *J. Theor. Biol.* 57:239–42.

Bell, G. 1978. The handicap principle in sexual selection. *Evolution* 32:872–85.

Cf. FISHER'S SEX RATIO THEORY; SEXUAL SELECTION

HANSTEIN'S THEORY [of histogenic layers] *see* **HISTOGEN THEORY**

HARD AND SOFT SELECTION [genetics]

Wallace, B. 1968. Polymorphism, population size, and genetic load. *Population Biology and Evolution.* Edited by R. C. Lewontin. Syracuse, N.Y.: Syracuse University Press, pp. 87–108.

______ . 1975. Hard and soft selection revisited. *Evolution* 29:465–73.

Christiansen, F. B. 1975. Hard and soft selection in a subdivided population. *Amer. Nat.* 109:11–16.

Cf. ALPHA SELECTION

HARDY-WEINBERG LAW [of gene:genotype frequencies]
= genetic equilibrium = SQUARE LAW = CASTLE'S LAW

Hardy, G. H. 1908. Mendelian proportions in mixed population. *Science* 28:49–50.

Weinberg, W. 1908. Über den Nachweis der Vererbung beim

Menschen. *J. Ver Vaterl. Naturk. Württ.* 64:368-82.

Stern, C. 1943. The Hardy-Weinberg law. *Science* 97:137–38.

———. 1962. Wilhelm Weinberg (1862-1937) biography. *Genetics* 47:1–5.

HARVEST OPTIMIZATION [ecology]

MacArthur, R. H. 1960. On the relation between reproductive value and optimal predation. *Proc. Natl. Acad. Sci.* 46:143–45.

Law, R. 1979. Harvest optimization in populations with age distributions. *Amer. Nat.* 114:250–59.

HASSELL-VARLEY MODEL [host-parasite interactions]

Hassell, M. P., & G. C. Varley. 1969. New inductive population model for insect parasites and its bearing on biological control. *Nature* 223:1133–37.

Allen, J. C. 1975. Mathematical models of species interactions in time and space. *Amer. Nat.* 109:319–42.

Cf. LOTKA-VOLTERRA MODEL; NICHOLSON-BAILEY MODEL; TIME-SPACE MODEL

HEADFUL HYPOTHESIS [of DNA content in phage head]

Streisenger, G., J. Emrich, & M. M. Stahl. 1967. Chromosome structure in phage T4. III. Terminal redundancy and length determination. *Proc. Natl. Acad. Sci.* 57:292–95.

Doermann, A. H. 1973. T4 and the rolling circle model of replication. *Annu. Rev. Genet.* 7:325–41.

HENNIG'S PRINCIPLE [taxonomy]

Hennig, W. 1966. *Phylogenetic Systematics.* Urbana: University of Illinois Press, pp. 73, 146.

Sneath, P. H. A., & R. R. Sokal. 1973. *Numerical Taxonomy.* San Francisco: Freeman.

HENNIG'S PROGRESSION RULE *see* **PROGRESSION RULE**

HENSEN'S THEORY *see* **PROTOPLASMIC BRIDGE THEORY**

HEREDITY CORRELATION, PRINCIPLE OF

Osborn, H. F. 1915. Origin of single characters as observed in fossil and living animals and plants. *Amer. Nat.* 59:193–239.

Thompson, D'A. W. 1942. *On Growth and Form.* 2d ed. Cambridge: Cambridge University Press, pp. 1019–20.

Osborn, H. F. 1915. Origin of single characters as observed in fossil and living animals and plants. *Amer. Nat.* 59:193–239.

Thompson, D'A. W. 1942. *On Growth and Form.* 2d ed. Cambridge: Cambridge University Press, pp. 1019–20.

HEREDITY, THEORIES OF

Weismann, A. 1902. *The Germ Plasm, A Theory of Heredity.* Transl. by W. Newton Parker & H. Rönnfeldt. New York: Scribner's.

Morgan, T. H. 1926. *The Theory of the Gene.* New Haven, Conn.: Yale University Press, chap. 2 (1964 reprint New York: Hafner.)

Grant, V. 1956. The development of a theory of heredity. *Amer. Sci.* 44:158–79.

Darlington, C. D. 1960. Chromosomes and the theory of heredity. *Nature* 187:892–95.

HERTWIG'S LAW or RULE [cell division]

Hertwig, O. 1884–85. Welchen Einfluss übt die Schwerkraft auf die Theilung der Zellen? *Jena. Z. Naturw.* 18:175–205.

Thompson, D'A. W. 1942. *On Growth and Form.* 2d ed. Cambridge: Cambridge University Press, p. 328.

Wilson, E. B. 1925. *The Cell in Development and Heredity.* 3d ed. New York: Macmillan, pp. 984–85.

Løvtrup, S. 1974. *Epigenetics.* New York: Wiley, pp. 128–29.

HETEROALLELIC REGENERATION *see* **GENE CONVERSION**

HETEROCHRONY [evolution; morphogenesis]

Goldschmidt, R. 1938. *Physiological Genetics.* New York: McGraw-Hill.

de Beer, G. R. 1958. *Embryos and Ancestors.* 3d ed. Oxford: Clarendon Press, chap. 5.

Bonner, J. T. 1965. *Size and Cycle.* Princeton, N.J.: Princeton University Press, pp. 119–20.

Gould, S. J. 1977. *Ontogeny and Phylogeny.* Cambridge: Belknap Press of Harvard University Press.

______. 1979. On the importance of heterochrony for evolutionary biology. *Syst. Zool.* 28:224–26.

Cf. RECAPITULATION

HETEROGENESIS [origin of life]

Breschet, G. 1823. Déviation organique. *Dictionnaire de médecine.* Paris: Bechet Jeune, 21 vols. 1821-28, 6:530.

Bulloch, W. 1938. *The History of Bacteriology*. London: Oxford University Press, chap. 4.

Farley, J. 1977. *The Spontaneous Generation Controversy from Descartes to Oparin*. Baltimore: Johns Hopkins Press.

Cf. ABIOGENESIS; SPONTANEOUS GENERATION

HETEROGONY [differential growth ratio]
= ALLOMETRY

Wilson, E.B. 1903. Notes on the reversal of asymmetry in the regeneration of the chelae in *Alpheus heterochelis*. *Biol. Bull.* 4:197–210.

Pezard, A. 1918. Le conditionnement physiologique des charactères sexuels secondaires chez les diseaux. *Bull. Biol. Fr. Belg.* 52:1–176.

Huxley, J. 1932. *Problems of Relative Growth*. London: Methuen.

Thompson, D'A. W. 1942. *On Growth and Form*. 2d ed. Cambridge: Cambridge University Press, pp. 206, 279.

HETEROLOGOUS ADVANTAGE HYPOTHESIS [of selection]

Fincham, J. 1972. Heterozygous advantage as a likely general basis for enzyme polymorphisms. *Heredity* 28:387–91.

Johnson, G. B. 1976. Genetic polymorphism and enzyme function. *Molecular Evolution*. Edited by F. J. Ayala. Sunderland, Mass.: Sinauer Associates, pp. 46–59.

HETEROSELECTION [genetics]

Carson, H. L. 1959. Genetic conditions which promote or retard the formation of species. *Cold Spr. Harb. Symp. Quant. Biol.* 24:87–105.

Crumpacker, D. W. 1967. Genetic loads in maize (*Zea mays* L) and other cross-fertilized plants and animals. *Evol. Biol.* 1:306–424.

Cf. HOMOSELECTION

HETEROSIS [genetics]
= hybrid vigor

East, E. M., & H. K. Hayes. 1912. Heterozygosis in evolution and in plant breeding. *U.S.D.A. Bur. Plant Industry Bulletin* 243:7–58.

Shull, G. H. 1914. Duplicate genes for capsule form in *Bursa bursa-pastoris*. *Z. Indukt. Abstammungs-Vererbungsl.* 12:97–149.

Jones, D. F. 1917. Dominance of linked factors as a means of accounting for heterosis. *Genetics* 2:466–79.

East, E. M. 1936. Heterosis. *Genetics* 21:375–97.

Ashby, E. 1937, Heterosis and the inheritance of quantitative characters. *Proc. Roy. Soc. Lond. B* 123:431–41.

Whaley, W. G. 1944. Heterosis. *Bot. Rev.* 10:461–98.

Crow, J. F. 1948. Alternative hypotheses of hybrid vigor. *Genetics* 33:477–87.

Schull, G. H. 1952. Beginnings of the heterosis concept. *Heterosis*. Edited by J. W. Gowen. Ames: Iowa State University Press, pp. 14–48.

Rendell, J. M. 1953. Heterosis. *Amer. Nat.* 87:129–38.

Berger, E. 1976. Heterosis and the maintenance of enzyme polymorphism. *Amer. Nat.* 110:823–39.

Cf. CONDITIONAL HETEROSIS

HETEROTROPH HYPOTHESIS [origin of life]

Hardin, G. 1949. *Biology: Its Human Implications*. San Francisco: Freeman.

______ . 1950. Darwin and the heterotroph hypothesis. *Sci. Mon.* 70:178–79.

Cf. ABIOGENESIS; AUTOTROPH HYPOTHESIS

HIERARCHICAL GENE CONTROL [population genetics]

Britten, R. J., & E. H. Davidson. 1969. Gene regulation for higher cells: a theory. *Science* 165:349–57.

Davidson, E. H., & R. J. Britten. 1973. Organization, transcription, and regulation in the animal genome. *Q. Rev. Biol.* 48:565–613.

Wallace, B. 1975. Gene control mechanisms and their possible bearing on the neutralist-selectionist controversy. *Evolution* 29:193–202.

HIERARCHICAL ORGANIZATION [of instinctive behavior]

Sherrington, C. 1906. *The Integrative Action of the Nervous System*. New York: Scribner's.

Tinbergen, N. 1950. The hierarchical organization of nervous mechanisms underlying instinctive behaviour. *Symp. Soc. Exp. Biol.* 4:305–12.

Dawkins, R. 1976. Hierarchical organization: a candidate principle for ethology. *Growing Points in Ethology*. Edited by P. P. G. Bateson & R. A. Hinde. Cambridge: Cambridge University Press, pp. 7–54.

Baerends, G. P. 1976. The functional organization of behavior. *Anim. Behav.* 24:726–38.

HISTOGEN THEORY [of apical meristems]

Hanstein, J. L. E. R. von. 1868. Die Scheitelzellgruppe im Vegetationspunkt der Phanerogamen. *Gratulationschrift Niederrheimischen Gesellschaft Natur- und Heilkunde.* Bonn: Marcus, pp. 109–34.

———. 1870. Die Entwickelung des Keimes der Monokotylen und der Dikotylen. *Bot. Abh.* 1:1–112.

Esau, K. 1977. *Anatomy of Seed Plants.* 2d ed. New York: Wiley, p. 228.

HIT THEORY [of radiation absorption]

Blau, M., & K. Altenburger. 1923. Über einige wirkungen von Strahlen II. Z. *Physik* 12:315–29.

Dessauer, F. 1923. Über einige wirkungen von Strahlen I. Z. *Physik* 12:38–47.

Hollaender, A. 1961. Hit and target theories. *Science* 134:1233.

Zimmer, K. G. 1966. The target theory. *Phage and the Origins of Molecular Biology.* Edited by J. Cairns, G. S. Stent, & J. D. Watson. Cold Spring Harbor, N.Y.: Cold Spring Harbor Laboratory of Quantitative Biology, pp. 33–42.

Cf. TARGET THEORY

HITCH-HIKING EFFECT [gene linkage]

Ohta, T., & M. Kimura. 1971. Behavior of neutral mutants influenced by associated overdominant loci in finite populations. *Genetics* 69:247–60.

———. 1976. Hitch-hiking effect—a counter reply. *Genet. Res.* 28:307–08.

Maynard Smith, J., & J. Haigh. 1974. The hitch-hiking effect of a favourable gene. *Genet. Res.* 23:23–35.

Cf. ASSOCIATIVE OVERDOMINANCE

HODGKIN-HUXLEY AXON [electrical properties of squid axon membrane]

= sodium theory

Hodgkin, A. L. 1951. The ionic basis of electrical activity in nerve and muscle. *Biol. Rev.* 26:339–409.

———, & A. F. Huxley. 1952. A quantitative description of membrane current and its application to conduction and excitation in nerve. *J. Physiol.* 117:500–44.

Cole, K. S. 1971. Some aspects of electrical studies of the squid giant axon membrane. *Biophysics and Physiology of Excitable Membranes.* Edited by W. J. Adelman, Jr. New York: Van Nostrand Reinhold, pp. 125–42.

HOFMEISTER'S RULE [leaf arrangement]

Hofmeister, W. 1868. Allgemeine Morphologie der Gewächse . . . *Handbuch der Physiologischen Botanik* . . . Leipzig: Engelmann.

Snow, M., & R. S. Snow. 1933. The interpretation of phyllotaxis. *Biol. Rev.* 9:132–37.

HOLISM [philosophy]

Smuts, J. C. 1926. *Holism and Evolution*. New York: Macmillan.

Phillips, J. 1935. Succession, development, the climax, and the complex organism: an analysis of concepts. III. The complex organism: conclusions. *J. Ecol.* 23:488–508.

Leppik, E. E. 1974. Phylogeny, hologeny and coenogeny, basic concepts of environmental biology. *Acta Biotheor.* 23:170–93.

HOLLOW CURVES [biogeography]

Willis, J. C. 1922. *Age and Area*. Cambridge: Cambridge University Press, chap. 18.

Chamberlin, T. C. 1924. Concerning the hollow curve of distribution. *Amer. Nat.* 58:350–74.

Mayr, E. 1942. *Systematics and the Origin of Species*. New York: Columbia University Press.

______ . 1963. *Animal Species and Evolution*. Cambridge: Belknap Press of Harvard University Press, p. 619.

Nicholson, R. J. 1951. A note on hollow curves. *New Phytol.* 50:138–39.

Cf. TYPOSTROPHE

HOLOGENESIS, THEORY OF [evolution]

Rosa, D. 1931. *L'Ologénèse*. Paris: Alcan.

Simpson, G. G. 1953. *The Major Features of Evolution*. New York: Columbia University Press, p. 304.

Nelson, G. 1973. Comments on Leon Croizat's biogeography. *Syst. Zool.* 22:312–20.

Baroni-Urbani, C. 1977. Hologenesis, phylogenetic systematics, and evolution. *Syst. Zool.* 26:343–46.

Cf. DIVERGENT DICHOTOMOUS MUTATION THEORY
PANBIOGEOGRAPHY

HOLOGENY [philosophy]

Leppik, E. E. 1957. Hologeny, a complementary concept of phylogeny. *Ann. Soc. Tartuentis Lund* (Sweden) Ser. Nov. 1:41–51.

______ . 1974. Phylogeny, hologeny and coenogeny, basic

concepts of environmental biology. *Acta Biotheor.* 23:170–93.

Cf. COENOGENY; HOLISM

HOLOPHYLESIS *see* **HOLOPHYLY**

HOLOPHYLY [evolution]

Ashlock, P. D. 1971. Monophyly and associated terms. *Syst. Zool.* 20:63–69.

———. 1972. Monophyly again. *Syst. Zool.* 21:430–38.

HOME RANGE [ethology]

Burt, W. H. 1943. Territoriality and home range concepts as applied to mammals. *J. Mammal.* 24:346–52.

Jewell, P. A. 1966. The concept of home range in mammals. *Symp. Zool. Soc. Lond.* 18:85–109.

Cf. TERRITORIALITY

HOMEORHESIS *see* **CANALIZATION**

HOMEOSTASIS [physiology]

Cannon, W. B. 1926. Physiological regulation of normal states: some tentative postulates concerning biological homeostasis. *Trans. Cong. Amer. Physicians and Surgeons* 12:31. (Reprinted in: *Selected Readings in the History of Physiology*. Edited by J. F. Fulton. Springfield, Ill.: Thomas, 1966, pp. 329–32.)

———. 1929. Organization for physiological homeostatics. *Physiol. Rev.* 9:399-431.

———. 1932. *The Wisdom of the Body*. New York: Norton.

Dobzhansky, T. 1955. A review of some fundamental concepts and problems of population genetics. *Cold Spr. Harb. Symp. Quant. Biol.* 20:1–15.

Society for Experimental Biology. 1964. *Homeostasis and Feedback Mechanisms. Symp. Soc. Exp. Biol.*, no. 18.

Slobodkin, L. B. 1968. Toward a predictive theory of evolution. *Population Biology and Evolution*. Edited by R. C. Lewontin. Syracuse, N.Y.: Syracuse University Press, pp. 187–205.

Langley, L. L., ed. 1973. *Homeostasis: Origins of the Concept.* Stroudsburg, Pa.: Dowden, Hutchinson & Ross.

Hardy, R. N. 1976. *Homeostasis*. London: Arnold.

Cf. BERNARD'S PRINCIPLE; CANALIZATION; DEVELOPMENTAL HOMEOSTASIS; EPIGENETIC HOMEOSTASIS; GENETIC HOMEOSTASIS; LE CHATELIER'S PRINCIPLE; MILIEU INTÉRIEUR; PHYSIOLOGICAL HOMEOSTASIS

 HOMOLOGOUS CHANGE, PRINCIPLE OF *see* **KARYOTYPIC ORTHOSELECTION**

HOMOLOGOUS SERIES or GENES, LAW OF [genetic variability]
= Vavilov's law

Vavilov, N. I. 1922. The law of homologous series in variation. *J. Genet.* 12:47–89.

______. 1951. The origin, variation, immunity, and breeding of cultivated plants. Transl. by K. S. Chester. *Chron. Bot.* 13:1–364.

Haldane, J. B. S. 1952. Variation. *New Biol.* 12:9–26.

Kupzow, A. J. 1975. Vavilov's law of homologous series at the fiftieth anniversary of its formulation. *Econ. Bot.* 29:372–79.

Cf. PARALLEL EVOLUTION

HOMOLOGOUS THEORY [of alternation of generations in plants]

Pringsheim, N. 1878. Über Sprassung der Moosfruchte und den Generationswechsel der Thallophyten. *Jahrb. Wiss. Bot.* 11:1–46.

Scott, D. H. 1896. Present position of morphological botany. *Report of the British Association of Liverpool,* pp. 992–1010.

Lang, W. H. 1909. Discussion on "alternation of generations" at the Linnean Society. *New Phytol.* 8:104–16.

Lal, M. 1961. *In vitro* production of apogamous sporogonia in *Physcomitrium coorgense* Broth. *Phytomorphology* 11:263–69.

Cf. ALTERNATION OF GENERATIONS; ANTITHETIC THEORY

HOMOLOGY [evolution]

Lankester, E. R. 1870. On the use of the term homology in modern zoology. *Ann. Mag. Nat. Hist.*, ser. 4, 6:34–43.

Bliakher, L. I. 1965 (publ. 1968). Evolution des notions "analogie" et "homologie." *Actes XIe Cong. Int. Hist. Sci.* 5:101–04.

Jardine, N. 1967. The concept of homology in biology. *Brit. J. Philos. Sci.* 18:125–39.

Bock, W. J. 1970. Discussion: the concept of homology. *Ann. N.Y. Acad. Sci.* 167:71–73.

Campbell, C. B. G., & W. Hodos. 1970. The concept of homology and the evolution of the nervous system. *Brain. Behav. Evol.* 3:353–67.

de Beer, G. 1971. Homology: an unsolved problem. *Oxford Ecology Readers*, II. Edited by F. F. Head & O. E. Lowenstein. Oxford: Oxford University Press.

Stebbins, G. L. 1974. *Flowering Plants: Evolution Above the Species Level.* Cambridge: Belknap Press of Harvard University Press, p. 142.

Hailman, J. P. 1976. Homology: logic, information, and efficiency. *Evolution, Brain, and Behavior: Persistent Problems.* Edited by R. B. Masterson, W. Hodos, & H. Jerison. Hillsdale, N.J.: Erlbaum, pp. 181–98.

HOMOPLASY [evolution]

Osborn, H. F. 1902. Homoplasy as a law of latent or potential homology. *Amer. Nat.* 36:259–71.

Cf. HOMOLOGY; PARALLEL EVOLUTION

HOMOSELECTION [genetics; evolution]

Carson, H. L. 1959. Genetic conditions which promote or retard the formation of species. *Cold Spr. Harb. Symp. Quant. Biol.* 24:87–105.

Richmond, R. C. 1972. Enzyme variability in the *Drosophila willistoni* group. III. Amounts of variability in the superspecies, *D. paulistorum. Genetics* 70:87–112.

Cf. HETEROSELECTION

HOPEFUL MONSTER [evolution]

Goldschmidt, R. B. 1933. Some aspects of evolution. *Science* 78:539–47.

———. 1940. *The Material Basis of Evolution.* New Haven, Conn.: Yale University Press, p. 390.

———. 1945. The structure of Podoptera, a homoeotic mutant of *Drosophila melanogaster. J. Morph.* 77:71–103.

Mayr, E. 1960. The emergence of evolutionary novelties. *Evolution After Darwin.* Edited by S. Tax. Chicago: University of Chicago Press, I:349–80.

Van Steenis, C. G. G. J. 1969. Plant speciation in Malesia, with special reference to the theory of non-adaptive saltatory evolution. *Biol. J. Linn. Soc.* 1:97–133.

Frazzetta, T. H. 1975. *Complex Adaptations in Evolving Populations.* Sunderland, Mass.: Sinauer Associates, chap. 4.

Cf. KRENKE'S RULE; SALTATION; MACROEVOLUTION

HOPKINS' BIOCLIMATIC LAW *see* **BIOCLIMATIC LAW**

HOPKINS' HOST-SELECTION PRINCIPLE [ecology; evolution]

Craighead, F. C. 1921. Hopkins' host-selection principle as related to certain cerambycid beetles. *J. Agric. Res.* 22:189–220.

Larson, A. D. 1927. The host-selection principle as applied to *Bruchus quadrimaculatus* Fab. *Ann. Entomol. Soc. Amer.* 20:37–80.

Thompson, W. R., & H. L. Parker. 1928. Host-selection in *Pyrausta nubilalis* Hübner. *Bull. Entomol. Res.* 18:359–64.

Dethier, V. G. 1954. Evolution of feeding preferences in phytophagous insects. *Evolution* 8:33–54.

Takata, N. 1961. Studies on the host preference of the common cabbage butterfly, *Pieris rapae crucivora* (Boisduval). XII. Successive rearing of the cabbage butterfly larva with certain host plants and its effect on the ovipositional preference of the adult. *Jap. J. Ecol.* 11:147–54.

Huettel, M. D., & G. L. Bush. 1972. The genetics of host selection and its bearing on sympatric speciation in *Procecidochares* (Diptera: Tephritidae). *Entomol. Exp. Appl.* 15:465–80.

HORMONAL CONCEPT [of seed dormancy]

Khan, A. A. 1971. Cytokinesis: permissive role in seed germination. *Science* 171:853–59.

______. 1977. Seed dormancy: changing concepts and theories. *The Physiology and Biochemistry of Seed Dormancy and Germination.* Edited by A. A. Khan. Amsterdam: North Holland Publishing, pp. 29–50.

HOROTELY [evolution]

= b-rates

Simpson, G. G. 1944. *Tempo and Mode in Evolution.* New York: Columbia University Press, chap. 4.

______. 1953. *The Major Features of Evolution.* New York: Columbia University Press, chap. 10.

Kurten, B. 1959. Rates of evolution in fossil mammals. *Cold Spr. Harb. Symp. Quant. Biol.* 24:205–15.

Cf. BRADYTELY; TACHYTELY

HULTEN'S HYPOTHESIS *see* **EQUIFORMAL PROGRESSIVE AREAS**

HUTCHINSON'S PARADOX *see* **PARADOX OF THE PLANKTON**

HYBRID SPECIATION [evolution]
Grant, V. 1971. *Plant Speciation*. New York: Columbia University Press, chap. 13.
Dobzhansky, T., F. J. Ayala, G. L. Stebbins, & J. W. Valentine. 1977. *Evolution* San Francisco: Freeman, pp. 220*ff*.

HYBRID-SUPERIORITY HYPOTHESIS [of vertebrate hybridization]
Moore, W. S. 1977. An evaluation of narrow hybrid zones in vertebrates. *Q. Rev. Biol.* 52:263–77.

HYBRID THEORY OF PARTHENOGENESIS [evolution]
Cuellar, O. 1974. On the origin of parthenogenesis in vertebrates: the cytogenetic factors. *Amer. Nat.* 108:625–48.

HYBRID VIGOR *see* **HETEROSIS**

HYBRID ZONES [evolution]
= INTERGRADATION ZONES = SUTURE-ZONES
Bigelow, R. S. 1965. Hybrid zones and reproductive isolation. *Evolution* 19:449–58.
Remington, C. L. 1968. Suture-zones of hybrid interaction between recently joined biotas. *Evol. Biol.* 2:321–438.
Gabow, S. A. 1975. Behavioral stabilization of a baboon hybrid zone. *Amer. Nat.* 109:701–12.
Woodruff, D. S. 1973. Natural hybridization and hybrid zones. *Syst. Zool.* 22:213–18.

HYBRIDIZATION OF THE HABITAT [ecology; evolution]
Anderson, E. 1948. Hybridization of the habitat. *Evolution* 2:1–9.
———. 1953. Introgressive hybridization. *Biol. Rev.* 28:280–307.
Heiser, C. B. 1973. Introgression re-examined. *Bot. Rev.* 39:347–66.

HYPERMUTATION THEORY *see* **SOMATIC MUTATIONAL HYPOTHESIS**

[I]

IDIOPLASM, THEORY OF [heredity]
Nägeli, C. W. von. 1884. *Mechanisch-physiologische Theorie*

der Abstammungslehre. München: Oldenbourg. (Transl. by V.A. Clark. 1914. *A Mechanico-physical Theory of Organic Evolution.*

Wilson, E. B. 1925. *The Cell in Development and Heredity*. 3d ed. New York: Macmillan, pp. 1037–39.

Wilkie, J. S. 1960 (publ. 1962). Nägeli's work on the fine structure of living matter. I & II. *Ann. Sci.* 16:11–42, 171–207.

Coleman, W. 1965. Cell nucleus and inheritance: an historical study. *Proc. Amer. Phil. Soc.* 109:124–58.

ILLEGITIMATE RECOMBINATION [genetics]

Campbell, A. 1962. Episomes. *Adv. Genet.* 11:101–45.

Franklin, N. 1971. Illegitimate recombination. *The Bacteriophage Lambda*. Edited by A. D. Hershey. Cold Spring Harbor, N.Y.: Cold Spring Laboratory of Quantitative Biology, pp. 175–94.

Starlinger, P. 1977. DNA rearrangements in procaryotes. *Annu. Rev. Genet.* 11:103–26.

Weisberg, R. A., & S. Adhya. 1977. Illegitimate recombination in bacteria and bacteriophage. *Annu. Rev. Genet.* 11:451–73.

IMMUNOLOGICAL INERTIA [of viviparity]

= immunological hypothesis

Anderson, J. M. 1965. Immunological inertia in pregnancy. *Nature* 206:786–87.

Beer, A. E., & R. E. Billingham. 1971. Immunobiology of mammalian reproduction. *Adv. Immunol.* 14:1–84.

Wilson, A. C., L. R. Maxson, & V. M. Sarich. 1974. Two types of molecular evolution. Evidence from studies of interspecific hybridization. *Proc. Natl. Acad. Sci.* 71:2843–47.

Cf. REGULATORY HYPOTHESIS

IMMUNOLOGICAL TOLERANCE [immunology]

Anderson, D., R. E. Billingham, G. H. Lampkin, & P. B. Medawar. 1951. The use of skin grafting to distinguish between monozygotic and dizygotic twins in cattle. *Heredity* 5:379–97.

Hašek, M., A. Lengerová, & T. Hraba. 1961. Transplantation immunity and tolerance. *Adv. Immunol.* 1:1–66.

Burnet, M. 1969. *Cellular Immunology*. London & New York: Cambridge University Press.

Howard, J. G., & N. A. Mitchison. 1975. Immunological tolerance. *Prog. Allergy* 18:43–96.

IMMUNOSURVEILLANCE HYPOTHESIS [immunology]
= immunological surveillance

Burnet, F. M. 1970. The concept of immunological surveillance. *Prog. Exp. Tumor Res.* 13:1–27.

——— . 1971. Immunological surveillance in neoplasia. *Transplant. Rev.* 7:3–25.

Essex, M. 1977. Immunity to leukemia, lymphoma, and fibrosarcoma in cats: a case for immunosurveillance. *Contemp. Top. Immunobiol.* 6:71–106.

IMPRINTING [learning in animals]

Heinroth, O. 1910. Beitrage zur Biologie, namentlich Ethologie und Physiologie der Anatiden. *Proc. (Verh.) 5th Int. Ornithol. Kong.* 5:589–702.

Lorenz, K. Z. 1935. Der Kumpan in der Umwelt des Vögels. *J. Ornithol.* 83:137–214. (Transl. by R. Martin. 1970. Companions as factors in the bird's environment. *Studies in Animal and Human Behaviour*. London: Methuen, I:101–258).

——— . 1937. The companion in the bird's world. *Auk* 54:245–73.

Hess, E. H. 1958. "Imprinting" in animals. *Sci. Amer.* 198(3): 81–90.

Sluckin, W. 1965. *Imprinting and Early Learning*. Chicago: Aldine.

Bateson, P. P. G. 1966. The characteristics and context of imprinting. *Biol. Rev.* 41:177–220.

Klopfer, P. H. 1973. *Behavioral Aspects of Ecology*. 2d ed. Englewood Cliffs, N.J.: Prentice-Hall, pp. 44–48.

INACTIVE-X HYPOTHESIS [genetics]
= single active x-chromosome hypothesis

Lyon, M. F. 1968. Chromosomal and subchromosomal inactivation. *Annu. Rev. Genet.* 2:31–52.

——— . 1972. X-chromosome inactivation and developmental patterns in mammals. *Biol. Rev.* 47:1–35.

——— . 1974. Mechanisms and evolutionary origins of variable x-chromosome activity in mammals. *Proc. Roy. Soc. Lond. B* 187:243–68.

Cattanach, B. M. 1975. Control of chromosome inactivation. *Annu. Rev. Genet.* 9:1–18.

INBREEDING [genctics]

Darwin, C. 1876. *The Effects of Cross- and Self-Fertilization in the Vegetable Kingdom*. London: Murray.

East, E., & D. F. Jones. 1919. *Inbreeding and Outbreeding: Their Genetic and Sociological Significance*. Philadelphia: Washington Square.

Fisher, R. A. 1949. *The Theory of Inbreeding*. Edinburgh & London: Oliver & Boyd.

INCIPIENT SPECIES [evolution]

Mayr, E. 1942. *Systematics and the Origin of Species*. New York: Columbia University Press, p. 155.

Dobzhansky, T. 1972. Species of *Drosophila:* new excitement in an old field. *Science* 177:664–69.

——— , & O. Pavlovsky. 1971. Experimentally created incipient species of *Drosophila*. *Nature* 230:289–92.

Cf. GEOGRAPHIC SPECIATION; SEMISPECIES

INCLUSIVE FITNESS, THEORY OF or HYPOTHESIS [evolution of behavior]

Hamilton, W. D. 1964. The genetical evolution of social behaviour. I & II. *J. Theor. Biol.* 7:1–16, 17–52.

——— . 1972. Altruism and related phenomena, mainly in social insects. *Annu. Rev. Ecol. Syst.* 3:193–232.

Kerr, W. E. 1969. Some aspects of the evolution of social bees (Apidae). *Evol. Biol.* 3:119–75.

Eberhard, M. J. W. 1975. The evolution of social behavior by kin selection. *Q. Rev. Biol.* 50:1–33.

Cf. KIN SELECTION

INDEPENDENCE PRINCIPLE [ionic movement in membranes; physiology]

Hodgkin, A. L., & A. F. Huxley. 1952. Currents carried by sodium and potassium ions through the membranes of the giant squid *Loligo*. *J. Physiol.* 116:449–72.

Hille, B. 1971. Voltage clamp studies on myelinated nerve fibers. *Biophysics and Physiology of Excitable Membranes*. Edited by W. J. Adelman, Jr. New York: Van Nostrand Reinhold, pp. 230–46.

INDEPENDENT ASSORTMENT, LAW OF *see* **MENDEL'S LAWS**

INDICATOR SPECIES [ecology]

Clements, F. E. 1928. *Plant Succession and Indicators*. New York: H. W. Wilson, pp. 209*ff*.

——— . 1920. Plant indicators: the relation of plant communities to process and practice. *Carnegie Inst. Wash. Publ.* 290.

INDIVIDUAL [ecology]

Jeuken, S. J. M. 1957. The concept "individual" in biology. *Acta Biotheor.* 10:57–86.

Janzen, D. H. 1977. What are dandelions and aphids? *Amer. Nat.* 111:586–89.

Cook, R. E. 1979. Asexual reproduction: further consideration. *Amer. Nat.* 113:769–72.

INDIVIDUALISTIC CONCEPT [of the plant association]
= Gleason's individualistic concept

Gleason, H. A. 1917. The structure and development of the plant association. *Bull. Torr. Bot. Club* 44:463–81.

______. 1926. The individualistic concept of the plant association. *Bull. Torr. Bot. Club* 53:7–26.

McIntosh, R. P. 1975. H. A. Gleason—"individualistic ecologist" 1882–1975: his contributions to ecological theory. *Bull. Torr. Bot. Club* 102:253–73.

Cf. CONTINUUM; GRADIENT ANALYSIS; SUPRAORGANISMIC CONCEPT

INDUSTRIAL MELANISM [evolution]

Kettlewell, H. B. D. 1961. The phenomenon of industrial melanism in lepidoptera. *Annu. Rev. Entomol.* 6:245–62.

Owen, D. F. 1961. Industrial melanism in North American moths. *Amer. Nat.* 95:227–33.

Creed, E. R. 1971. Industrial melanism in the two-spot ladybird and smoke abatement. *Evolution* 25:290–93.

INERTIA [of ecosystems]

Orians, G. H. 1975. Diversity, stability and maturity in natural ecosystems. *Unifying Concepts in Ecology*. Edited by W. H. Van Dobben & R. H. Lowe-McConnel. The Hague: Junk, pp. 64–65.

Westman, W. E. 1978. Measuring the inertia and resilience of ecosystems. *Bioscience* 28:705–10.

INFECTIVE PRINCIPLE [in evolution toward inbreeding]

Stebbins, G. L. 1957. Self-fertilization and population variability in the higher plants. *Amer. Nat.* 41:337–54.

Jain, S. K. 1976. The evolution of inbreeding in plants. *Annu. Rev. Ecol. Syst.* 7:469–95.

INHIBITOR CONCEPT [of seed dormancy]

Nutile, G. E. 1945. Inducing dormancy in lettuce with coumarin. *Plant Physiol.* 20:433–42.

Khan, A. A. 1977. Seed dormancy: changing concepts and theories. *The Physiology and Biochemistry of Seed Dormancy and Germination*. Edited by A. A. Khan. Amsterdam: North Holland Publishing, pp. 29–50.

INHIBITOR-PROMOTER CONCEPT [of seed dormancy]

Amen, R. D. 1968. A model of seed dormancy. *Bot. Rev.* 34:1–30.

Khan, A. A. 1977. Seed dormancy: changing concepts and theories. *The Physiology and Biochemistry of Seed Dormancy and Germination.* Edited by A. A. Khan. Amsterdam: North Holland Publishing, pp. 29–50.

INNATE RELEASING MECHANISM [behavior]

Lorenz, K. 1935. Der Kumpan in der Umwelt des Vögels. *J. Ornithol.* 83:137–214.

______ . 1950. The comparative method in studying innate behavior patterns. *Symp. Soc. Exp. Biol.* 4:221–68.

Thorpe, W. H. 1956. *Learning and Instinct in Animals.* London: Methuen.

Klopfer, P. H. 1973. *Behavioral Aspects of Ecology.* Englewood Cliffs, N.J.: Prentice-Hall, pp. 123–29.

INSECT-PLANT COEVOLUTION THEORY

Opler, P. A. 1974. Oaks as evolutionary islands for leaf-mining insects. *Amer. Sci.* 62:67–73.

Claridge, M. F., & M. R. Wilson. 1978. British insects and trees: a study in island biogeography or insect/plant coevolution. *Amer. Nat.* 112:451–56.

INSIDE-OUTSIDE THEORY [development]

Tarkowski, A. K., & J. Wroblewska. 1967. Development of blastomeres of mouse eggs isolated at the 4- and 8-cell stage. *J. Embryol. Exp. Morphol.* 18:155-80.

Johnson, M. H., A. H. Handyside, & P. Braude. 1977. Control mechanisms in early mammalian development. *Development in Mammals.* Edited by M. H. Johnson. Amsterdam: North Holland Publishing, vol. 2. pp. 67-97.

Bratt, H. P. M. 1978. Lipids and transitions in embryos. *Development in Mammals.* Vol. 3. Edited by M. H. Johnson. Amsterdam: North Holland Publishing, pp. 83–129.

INSTANT SPECIATION [evolution]

Bush, G. L. 1975. Modes of animal speciation. *Ann. Rev. Ecol. Syst.* 6:339–364.

______ , S. M. Case, A. C. Wilson, & J. L. Patton. 1977. Rapid speciation and chromosomal evolution in mammals. *Proc. Natl. Acad. Sci.* 74:3942–46.

Hamilton, W. D. 1978. Evolution and diversity under bark. *Symp. Roy. Ent. Soc. Lond.* 9:154–175.

Cf. FOUNDER EFFECT; GENETIC DRIFT

INSTINCT [ethology]

Lorenz, K. 1937. The establishment of the instinct concept. (Transl. by R. Martin. 1970. *Studies in Animal and Human Behaviour*. London: Methuen, I:259–315).

Tinbergen, N. 1951. *The Study of Instinct*. Oxford: Clarendon Press.

INTEGRATIVE ACTION [neurophysiology]

Sherrington, C. 1906. *The Integrative Action of the Nervous System*. New York: Scribner's.

Swazey, J. P. 1968. Sherrington's concept of integrative action. *J. Hist. Biol.* 1:57–89.

———. 1969. *Reflexes and Motor Integration: Sherrington's Concept of Integrative Action*. Cambridge: Harvard University Press.

INTERDEME SELECTION [evolution]

Wright, S. 1931. Evolution in Mendelian populations. *Genetics* 16:97–159.

Gilmour, J. S. L., & J. W. Gregor. 1939. Demes: a suggested new terminology. *Nature* 144:333.

Wright, S. 1956. Modes of selection. *Amer. Nat.* 90:5–24.

Levin, B. R., & W. L. Kilmer. 1974. Interdemic selection and the evolution of altruism: a computer simulation study. *Evolution* 28:527–45.

Cf. DEME

INTERFERENCE [genetics]

= chiasma interference = chromatid interference = chromosomal interference

Muller, H. J. 1916. The mechanism of crossing-over. I-IV. *Amer. Nat.* 50:193–221, 284–305, 350–66, 421–34.

Mather, D. 1933. The relation between chiasmata and crossing-over in diploid and triploid *Drosophila melanogaster*. *J. Genet.* 27:243–59.

Haldane, J. B. S. 1931. The cytological basis of genetical interference. *Cytologia* 3:54–65.

INTERGRADATION ZONES, PRIMARY AND SECONDARY [biogeography; evolution]

= HYBRID ZONES = SUTURE ZONES

Mayr, E. 1942. *Systematics and the Origin of Species*. New York: Columbia University Press, p. 99.

———. 1963. *Animal Species and Evolution*. Cambridge:

Belknap Press of Harvard University Press, chap. 13.

Gabow, S. A. 1975. Behavioral stabilization of a baboon hybrid zone. *Amer. Nat.* 109:701–12.

INTERMEDIATE DISTURBANCE HYPOTHESIS [of species diversity]

Grime, J.P. 1973. Control of species density in herbaceous vegetation. *J. Environ. Manage.* :151–67.

Connell, J. H. 1978. Diversity in tropical rainforests and coral reefs. *Science* 199:1302–10.

Huston, M. 1979. A general hypothesis of species diversity. *Amer. Nat.* 113:81–101.

Fox, J. F. 1979. Intermediate disturbance hypothesis. *Science* 204:1344–45.

INTRACELLULAR PANGENESIS [heredity]

DeVries, H. 1889. *Intracellulare Pangenesis*. Jena: Fischer. (Transl. by C. S. Gager. 1910. Chicago: Open Court.)

Portugal, F. H., & J. S. Cohen. 1977. *A Century of DNA*. Cambridge: M. I. T. Press, chap. 5.

INTRAGENERIC ISOLATION *see* **ECOLOGICAL ISOLATION**

INTRAGENIC RECOMBINATION *see* **GENE CONVERSION**

INTRINSIC RATE OF NATURAL INCREASE *see* **BIOTIC POTENTIAL**

INTROGRESSIVE HYBRIDIZATION [ecology; evolution]

Anderson, E. 1949. *Introgressive Hybridization*. New York: Wiley.

______ . 1953. Introgressive hybridization. *Biol. Rev.* 28:280–307.

______ , & L. Hubricht. 1938. Hybridization in *Tradescantia*. III. The evidence for introgressive hybridization. *Amer. J. Bot.* 25:396–402.

Heiser, C. B., Jr. 1973. Introgression re-examined. *Bot. Rev.* 39:347–66.

INVAGINATION THEORY [of mitochondrion origin]

Raff, R.A., & H.R. Mahler. 1972. The non-symbiotic origin of mitochondria. *Science* 177:575–82.

INVERSE PYRAMID OF HABITATS *see* **PYRAMID OF NUMBERS**

INVERSE-SQUARE LAW [ratio of sound intensity and distance]

Peterson, A. P. G. & E. E. Gross. 1967. *Handbook of Noise Measurement*. West Concord, Mass.: General Radio.

Morton, E. S. 1975. Ecological sources of selection on avian sounds. *Amer. Nat.* 109:17–34.

IONIC THEORY *see* **BERNSTEIN MEMBRANE THEORY**

IRREVERSIBILITY, LAW OF [evolution]

Arber, A. 1919. On atavism and the law of irreversibility. *Amer. J. Sci.* 48:27–32.

Stebbins, G. L. 1950. *Variation and Evolution in Plants*. New York: Columbia University Press, p. 497.

ISLAND BIOGEOGRAPHY, THEORY OF

= EQUILIBRIUM THEORY

Preston, F. W. 1962. The canonical distribution of commonness and rarity. *Ecology* 43:185–215; 410–32.

MacArthur, R. H., & E. O. Wilson. 1963. An equilibrium theory of insular zoogeography. *Evolution* 17:373–87.

______. 1967. *The Theory of Island Biogeography*. Princeton: Princeton University Press.

Roughgarden, J. 1979. *Theory of Population Genetics and Evolutionary Ecology: An Introduction*. New York: Macmillan, chap. 24.

Cf. NONEQUILIBRIUM HYPOTHESIS; SPECIES-AREA CURVE

ISOALLELIC HYPOTHESIS [evolution]

Selander, R. K. 1970. Biochemical polymorphism in populations of the house mouse and old-field mouse. *Symp. Zool. Soc. Lond.* 26:73–91.

Cf. NEUTRAL THEORY

ISOLATING MECHANISMS [evolution]

Du Rietz, G. E. 1930. The fundamental units of botanical taxonomy. *Svensk Bot. Tidskr.* 24:333–428.

Dobzhansky, T. 1937. *Genetics and the Origin of Species*. New York: Columbia University Press, chap. 7.

______. ed. 1942. Symposium on Isolating Mechanisms. *Biol. Symp.* 6:217–87.

Mayr, E. 1963. *Animal Species and Evolution*. Cambridge: Belknap Press of Harvard University Press, chap. 5.

Levin, D. A. 1978. The origin of isolating mechanisms in

flowering plants. *Evol. Biol.* 11:185–317.
Cf. REPRODUCTIVE ISOLATION

ISOLATION [in evolution] *see* **ECOLOGICAL ISOLATION; GEOGRAPHIC ISOLATION; ISOLATION BY DISTANCE; REPRODUCTIVE ISOLATION**

ISOLATION BY DISTANCE [genetics; evolution]

Wright, S. 1943. Isolation by distance. *Genetics* 28:114–38.

______ . 1946. Isolation by distance under diverse systems of mating. *Genetics* 31:39–59.

Rohlf, F. J., & G. D. Schnell. 1971. An investigation of the isolation-by-distance model. *Amer. Nat.* 105:295–324.

Felsenstein, J. 1975. A pain in the torus: some difficulties with models of isolation by distance. *Amer. Nat.* 109:359–68.

Cf. GENETIC DISTANCE

[J]

JACOB-MONOD OPERON THEORY *see* **OPERON THEORY**

JAMIN'S CHAINS [theory of sap ascent]

Zimmermann, M. H. 1974. Long distance transport. *Plant Physiol.* 54:472–79.

JANICKI RULE [parasite ontogeny]

Janiszewska, J. 1949. Parasitogenetic rules. Janicki rule. *Zool. Pol.* 5:31–34.

Szidat, L. 1956. Geschichte, anwendung und einige folgerungen aus den parasitogenetischen regeln. Z. *Parasitenk.* 17:237–68.

JORDAN'S LAW [of geographic speciation]
= law of geminate species = Wagner's law

Allen, J. A. 1877. The influence of physical conditions in the genesis of species. *Radical Review* 1:108–40.

______ . 1905. The evolution of species through climatic conditions. *Science* 22:661-68.

Jordan, D. S. 1905. The origin of species through isolation. *Science* 22:545–62.

______ . 1906. Concerning variation in animals and plants. *Pop. Sci. Mon.* 69:481–502.

______ . 1908. The law of geminate species. *Amer. Nat.* 42:73–80.

———. 1926. Isolation with segregation as a factor in organic evolution. *Smithson. Inst. Annu. Rep.* 1925:321–26.

———. 1928. Evolution—its meaning. *Creation by Evolution*. Edited by F. Mason. New York: Macmillan, pp. 1–12.

Cain, S. A. 1944. *Foundations of Plant Geography*. New York: Harper, chap. 23.

Mayr, E. 1963. *Animal Species and Evolution*. Cambridge: Belknap Press of Harvard University Press, p. 487.

JUMPING GENES [genetics]

Kolata, G. B. 1976. Jumping genes: a common occurrence in cells. *Science* 193:392–94.

[K]

KARYOTYPIC ORTHOSELECTION [evolution]
= orthoselection

Simpson, G. G. 1944. *Tempo and Mode in Evolution*. New York: Columbia University Press.

White, M. J. D. 1962. Genetic adaptation. *Aust. J. Sci.* 25:179–86.

———. 1973. *Animal Cytology and Evolution*. 3d ed. Cambridge: Cambridge University Press, chap. 12.

Mayr, E. 1976. *Evolution and the Diversity of Life*. Cambridge: Belknap Press of Harvard University Press, p. 50.

KEW RULE [botanical nomenclature]

Lawrence, G. H. M. 1951. *Taxonomy of vascular plants*. New York: Macmillan, p. 197.

KIN SELECTION, THEORY OF [behavior; evolution]
= KINSHIP THEORY

Snell, G. D. 1932. The role of male parthenogenesis in the evolution of the social Hymenoptera. *Amer. Nat.* 66:381–84.

Hamilton, W. D. 1964. The genetical evolution of social behaviour. I & II. *J. Theor. Biol.* 7:1–16, 17–52.

———. 1972. Altruism and related phenomena, mainly in social insects. *Annu. Rev. Ecol. Syst.* 3:193–232.

Maynard Smith, J. 1964. Kin selection and group selection. *Nature* 201:1145–47.

Wilson, E. O. 1971. *The Insect Societies*. Cambridge: Harvard University Press, chap. 17.

______. 1975. *Sociobiology: The New Synthesis*. Cambridge: Harvard University Press.

Brown, J. L. 1974. Alternate routes to sociality in jays—with a theory for the evolution of altruism and communal breeding. *Amer. Zool.* 14:63–80.

______. 1978. Avian communal breeding systems. *Annu. Rev. Ecol. Syst.* 9:123–55.

Eberhard, M. J. W. 1975. The evolution of social behavior by kin selection. *Q. Rev. Biol.* 50:1–33.

Sahlins, M. 1976. *The Use and Abuse of Biology: An Anthropological Critique of Sociobiology*. Ann Arbor: University of Michigan Press.

Evans, H. E. 1977. Extrinsic versus intrinsic factors in the evolution of insect sociality. *Bioscience* 27:613–17.

Schulman, S. R. 1978. Kin selection, reciprocal altruism, and the principle of maximization: a reply to Sahlins. *Q. Rev. Biol.* 53:283–86.

Roughgarden, J. 1979. *Theory of Population Genetics and Evolutionary Ecology: An Introduction*. New York: Macmillan, chap. 14.

Cf. GROUP SELECTION

KINSHIP THEORY [behavior; evolution]

= KIN SELECTION

Hamilton, W. D. 1963. The evolution of altruistic behavior. *Amer. Nat.* 97:354–56.

Trivers, R. L., & H. Hare. 1976. Haplodiploidy and the evolution of the social insects. *Science* 191:249–63.

KLADOGENESIS *see* **CLADOGENESIS**

KLEIBER'S LAW [animal energetics]

Kleiber, M. 1932. Body size and metabolism. *Hilgardia* 6:315–53.

______. 1961. *The Fire of Life*. New York: Wiley, chap. 10.

Janis, C. 1976. The evolutionary strategy of the Equidae and the origins of rumen and cecal digestion. *Evolution* 30:757–74.

KLUGE-KERFOOT PHENOMENON [genetics]

Kluge, A. G., & W. C. Kerfoot. 1973. The predictability and regularity of character divergence. *Amer. Nat.* 107:426–42.

Sokal, R. R. 1976. The Kluge-Kerfoot phenomenon reexamined. *Amer. Nat.* 110:1077–91.

Pierce, B. A., & J. B. Mitton. 1979. A relationship of genetic variation within and among populations: an extension of the

Kluge–Kerfoot phenomenon. *Syst. Zool.* 28:63–70.

KNIGHT-DARWIN LAW ["nature abhors inbreeding"]

Knight, A. 1799. An account of some experiments on the fecundation of vegetables. *Phil. Trans. Roy. Soc. Lond.* 89:195–204.

Darwin, C. 1858. On the agency of bees in the fertilization of papilionaceous flowers, and on the crossing of kidney beans. *Ann. Mag. Nat. Hist.*, ser. 3, 2:459–65.

Darwin, F. 1898. The Knight-Darwin law. *Nature* 58:630–32.

Ghiselin, M. T. 1974. *The Economy of Nature and the Evolution of Sex.* Berkeley: University of California Press, p. 117.

KOLMOGOROV'S THEOREM [ecology]

Kolmogorov, A. 1936. Sulla teoria di Volterra della lotta per l'esistenza. *Giornale Inst. Ital. Attuari* 7:74–80.

Bulmer, M. G. 1976. The theory of prey-predator oscillations. *Theor. Pop. Biol.* 9:137–50.

KÖRPER-KAPPE CONCEPT [of the apical meristem] *see* **BODY-CAP CONCEPT**

KREBS EFFECT [population density]

Krebs, C., B. Keller, & R. Tamarin. 1969. *Microtus* population biology. *Ecology* 50:587–607.

MacArthur, R. H. 1972. *Geographical Ecology.* New York: Harper & Row, p. 118.

KRENKE'S RULE [evolution]

Krenke, N. P. 1933–35. Somatische Indikatoren und Factoren der Formbildung. *Phänogenetische Variabilität, Abh. Abt. Phytomorphogenese*, Inst. für Biol. Moscow. Band I:11–415. (In Russian, German summary.)

Van Steenis, C. G. G. J. 1969. Plant speciation in Malesia with reference to the theory of non-adaptive, saltatory evolution. *Biol. J. Linn. Soc.* 1:97–133.

Meyen, S. V. 1973. Plant morphology in its nomothetical aspects. *Bot. Rev.* 39:205-60.

KROGH'S FORMULA or LAW [relating growth to temperature]

Krogh, A. 1914. On the influence of temperature on the rate of embryonic development. *Z. Allg. Physiol.* 16:163–90.

Allee, W. C., & K. P. Schmidt. 1951. *Ecological Animal Geography.* 2d ed., rev. by R. Hesse. New York: Wiley, p. 20.

LACK'S PRINCIPLE OR HYPOTHESIS [of reproductive effectiveness]

Lack, D. 1954. The evolution of reproductive rates. *Evolution as a Process*. Edited by J. S. Huxley, A. C. Hardy, & E. B. Ford. London: Allen & Unwin, pp. 143–56.

Williams, G. C. 1966. Natural selection, the costs of reproduction, and a refinement of Lack's principle. *Amer. Nat.* 100:687–90.

Payne, R. B. 1974. The evolution of clutch size and reproductive rates in parasitic cuckoos. *Evolution* 28:169–81.

Stearns, S. C. 1976. Life-history tactics: a review of the ideas. *Q. Rev. Biol.* 51:3–47.

LAG LOAD [evolution]

Maynard Smith, J. 1976. What determines the rate of evolution? *Amer. Nat.* 110:331–38.

LAGRANGE'S THEORY [physiology]

Hassenfratz, J. 1791. Mémoire sur la combinaison de l'oxigène avec le carbone et l'hydrogène du sang, et sur la dissolution de l'oxigène dans le sang, et sur la manière dont le calorique se dégage. *Ann. Chim.* 9:266–67.

Hall, D. L. 1971. The iatromechanical background of Lagrange's theory of animal heat. *J. Hist. Biol.* 4:245–48.

LAMARCKIAN EVOLUTION

= autogenesis = Lamarckism

Lamarck, J. B. de. 1809. *Philosophie Zoologique*. . . .2 vols. Paris: Bailliere.

Hutten, F. W. 1899. *Darwinism and Lamarckism, Old and New: Four Lectures*. New York & London: Putnam's.

Packard, A. S. 1901. *Lamarck, the Founder of Evolution: His Life and Work*. New York: Longmans.

Crombie, A. C. 1953. The idea of organic evolution. *Discovery* March, 1953, pp. 92–97.

Gillespie, C. C. 1956. The formation of Lamarck's evolutionary theory. *Arch. Int. Hist. Sci.* 35:323–38.

Cannon, H. G. 1959. *Lamarck and Modern Genetics*. Manchester: Manchester University Press.

Darlington, C. D. 1964. *Genetics and Man*. New York: Macmillan, chap. 11.

Mayr, E. 1972. Lamarck revisited. *J. Hist. Biol.* 5:55–94.

Grasse, P. 1973. *L'evolution du Vivant*. Paris: Albin Michel.

Cf. LYSENKOISM; MICHURINISM

LAMARCK'S LAWS *see* **LAMARCKIAN EVOLUTION**

LAMINAR THEORY [of the tubular leaf of *Nepenthes]*

Wunschmann, E. 1872. Ueber die Gattung Nepenthes. Dissertation, Berlin University, 46p.

Troll, W. 1939. *Vergleichende Morphologie der höheren Pflanzen*. Band I. *Vegetationsorgane*, teil 2. Berlin: Gebrüder Borntraeger.

Franck, D. H. 1976. The morphological interpretation of epiascidate leaves. *Bot. Rev.* 42:345–88.

LAND-BRIDGE [biogeography]

Schuchert, C. 1932. Gondwana land bridges. *Bull. Geol. Soc. Amer.* 43:875–915.

Croizat, L. 1958. *Panbiogeography*. Caracas: By the Author, vol. I, p. 91; chap. 8.

Steenis, C. G. G. J. van. 1962. The land-bridge theory in botany with particular reference to tropical plants. *Blumea* 11:235–372.

Good, R. 1974. *The Geography of the Flowering Plants*. 4th ed. London: Longman, chap. 21.

Fichman, M. 1976. Wallace: zoogeography and the problem of land bridges. *J. Hist. Biol.* 10:45–63.

LANSING EFFECT [physiology]

Lansing, A. I. 1947. A transmissible, cumulative and reversible factor in aging. *J. Gerontol.* 2:228–39.

Lints, F. A., & C. Hoste. 1974. The Lansing effect revisited. I. Life-span. *Exp. Gerontol.* 9:51–69.

———. 1977. The Lansing effect revisited. II. Cumulative and spontaneously reversible parental age effects on fecundity in *Drosophila melanogaster*. *Evolution* 31:387–404.

LARGESSE OF THE GENOME [evolution of mimicry]

Turner, J. R. G. 1977. Butterfly mimicry: the genetical evolution of an adaptation. *Evol. Biol.* 10:163–206.

LAST MAN ON THE RIGHT PROBLEM [of development]

Wilde, C. E. 1974. Time flow in differentiation and morphogenesis. *Concepts of Development*. Edited by J. Lash & J. R. Whittaker. Stamford, Conn.: Sinauer Associates, pp. 241–60.

LEAF-SKIN THEORY [of the stem]

Saunders, E. R. 1922. The leaf-skin theory of the stem. *Ann. Bot.* 36:135–65.

Neubauer, H. F. 1971. Über Blattgrund und Stammknoten von Pelargonien. *Österr. Bot.* Z. 119:141–53.

LEAP-FROG SPECIATION [evolution]

Dodson, C. H. 1962. The importance of pollination in the evolution of the orchids of tropical America. *Amer. Orchid Soc. Bull.* 31:525–34, 641–49, 731–35.

Pijl, L. van der, & C. H. Dodson. 1967. *Orchid Flowers, Their Pollination and Evolution.* Coral Gables, Fla.: University of Miami Press, chap. 13.

LEAST-NAVIGATION HYPOTHESIS [selection for efficiency in navigation]

Baker, R. R. 1978. *The Evolutionary Ecology of Animal Migration.* New York: Holmes & Meier, p. 861.

LE CHATELIER'S PRINCIPLE [physiology; biochemistry]

Lotka, A. J. 1925. *Elements of Physical Biology.* Baltimore: Williams & Wilkins, pp. 281*ff*.

Cannon, H. G. 1958. *The Evolution of Living Things.* Manchester: Manchester University Press, p. 158.

Wardlaw, C. W. 1965. *Organization and Evolution in Plants.* London: Longmans, pp. 420–22.

Cf. HOMEOSTASIS

LEPIDOMORIAL THEORY [of tooth origin]

Ørvig, T. 1951. Histologic studies of Placoderms and fossil elasmobranchs. I. The exoskeleton, with remarks on the hard tissues of lower vertebrates in general. *Ark. Zool.*, ser. 2, II, no. 2.

Stensiö, E. A. 1961. *Permian Vertebrates.* Geology of the Arctic. Toronto: University of Toronto Press.

Peyer, B. 1968. *Comparative Odontology.* Chicago: University of Chicago Press, p. 43.

LEQUESNE'S UNIQUELY DERIVED CHARACTER CONCEPT *see* **UNIQUELY DERIVED CHARACTER CONCEPT**

LETHAL-HIT HYPOTHESIS [of the origin of complete mutations]

Vielmetter, W., & C. M. Wiedner. 1959. Mutagens und inaktivierende Wirkung salpetriger Säure auf freie Partikel des Phagen T2. *Z. Naturforsch. B.* 15:312–17.

Nasim, A., & C. Auerbach. 1967. The origin of complete and mosaic mutants from mutagenic treatment of single cells. *Mutat. Res.* 4:1–14.

Auerbach, C., & B. J. Kilbey. 1971. Mutation in eukaryotes. *Annu. Rev. Genet.* 5:163–218.

Cf. DUAL-MECHANISM HYPOTHESIS; MASTER-STRAND HYPOTHESIS; REPAIR HYPOTHESIS

LETHAL MUTATION THEORY [origin of genetic code]

Sonneborn, T. M. 1965. Degeneracy of the genetic code: extent, nature, and genetic implications. *Evolving Genes and Proteins.* Edited by V. Bryson & H. Vogel. New York: Academic Press, pp. 377–97.

Fitch, W. M. 1966. The relation between frequencies of amino acids and ordered trinucleotides. *J. Mol. Biol.* 16:1–8.

——— . 1973. Aspects of molecular evolution. *Annu. Rev. Genet.* 7:343–80.

Woese, C. 1969. Models for the evolution of codon assignments. *J. Mol. Biol.* 43:235–40.

LEVELS OF ORGANIZATION [in biological systems]

Rowe, J. S. 1961. The level-of-integration concept and ecology. *Ecology* 42:420–27.

Guttman, B. S. 1976. Is "levels of organization" a useful biological concept? *Bioscience* 26:112–13.

MacMahon, J. A., D. L. Phillips, J. V. Robinson, & D. J. Schimpf. 1978. Levels of biological organization: an organism-centered approach. *Bioscience* 28:700–04.

LEVINS' THEORY OF FITNESS *see* **FITNESS** [genetic]

LEWIS EFFECT [genetics]

= cis-trans effect

Pontecorvo, G. 1955. Gene structure and action in relation to heterosis. *Proc. Roy. Soc. Lond. B* 144:171–77.

LIEBIG'S LAW OF THE MINIMUM *see* **MINIMUM, LAW OF**

LIFE FORMS [ecology]

Humboldt, A. von. & A. Bonpland. 1805. *Essai sur la géographie des plantes.* Paris: Levrault, Schoell.

Clements, F. E. 1928. *Plant Succession and Indicators.* New York: H. W. Wilson.

Raunkiaer, C. 1934. *The Life Forms of Plants and Statistical*

Plant Geography. Oxford: Clarendon Press.

Dansereau, P. 1951. Description and recording of vegetation upon a structural basis. *Ecology* 32:172–229.

Böcher, T. W. 1977. Convergence as an evolutionary process. *Bot. J. Linn. Soc.* 75:1–19.

LIFE HISTORY THEORY [evolution]

Cole, L. C. 1954. The population consequences of life history phenomena. *Q. Rev. Biol.* 29:103–37.

Gadgill, M., & W. H. Bossert. 1970. Life history consequences of natural selection. *Amer. Nat.* 104:1–24.

Stearns, S. C. 1976. Life-history tactics: a review of the ideas. *Q. Rev. Biol.* 51:3–47.

______. 1977. The evolution of life history traits: a critique of the theory and a review of the data. *Ann. Rev. Ecol. Syst.* 8:145–71.

Cf. COLE'S RESULT

LIFE ZONES [biogeography]

Merriam, C. H. 1890. Results of a biological survey of the San Francisco mountain region and desert of the Little Colorado, Arizona. *North Amer. Fauna* 3:1–136.

______. 1898. Life zones and crop zones of the United States. *U.S.D.A. Bur. Biol. Survey Bull.* 10:1–73.

Holdridge, L. R. 1967. *Life Zone Ecology*. San Jose, Costa Rica: Tropical Science Center.

MacArthur, R. H. 1972. *Geographical Ecology*. New York: Harper & Row, pp. 159*ff*.

LIMITED REALIZATION, PRINCIPLE OF *see* **SPIEGELMAN'S RULE**

LIMITING FACTORS, LAW OF *see* **MINIMUM, LAW OF**

LINEARIZATION HYPOTHESIS [ecology; evolution]

Patten, B. C. 1975. Ecosystem linearization: an evolutionary design problem. *Amer. Nat.* 109:529–39.

LINES OF DEFENSE [by prey against predators]

Kettlewell, H. B. D. 1959. Brazilian insect adaptations. *Endeavour* 18:200–10.

Robinson, M. H. 1969. Defenses against visually hunting predators. *Evol. Biol.* 3:225–59.

LINKAGE DISEQUILIBRIUM *see* **GAMETIC DISEQUILIBRIUM**

LINKAGE THEORY [genetics]

Bateson, W., & R. C. Punnett. 1906. Experimental studies in the physiology of heredity. *Rep. Evol. Comm. Roy. Soc. Lond.* 3:1–131.

Morgan, T. H., A. H. Sturtevant, H. J. Muller, to C. B. Bridges. 1923. *The Mechanism of Mendalian Heredity.* Rev. ed. New York: Holt, chap. 3.

Owen, A. R. G. 1950. The theory of genetical recombination. *Adv. Genet.* 3:117–57.

Sturtevant, A. H. 1965. *A History of Genetics.* New York: Harper & Row, chap. 6.

LOCALE ODOR THEORY [of bee recruitment]

Wenner, A. M. 1971. *The Bee Language Controversy.* Boulder, Colo.: Educational Programs Improvement.

Gould, J. L. 1976. The dance-language controversy. *Q. Rev. Biol.* 51:211–44.

LONG-DISTANCE DISPERSAL [biogeography]

Baker, H. G. 1955. Self-compatibility and establishment after "long-distance" dispersal. *Evolution* 9:347–49.

Carlquist, S. 1965. *Island Life.* New York: Natural History Press.

———. 1966. The biota of long-distance dispersal. I. Principles of dispersal and evolution. *Q. Rev. Biol.* 41:247–70.

Cf. BAKER'S LAW

"LOSING YOUR MARBLES THEORY" [of predation]

Smith, F. E. 1972.Spatial heterogeneity,stability, and diversity in ecosystems. *Trans. Conn. Acad. Arts Sci.* 44:309–35.

Charnov, E. L., G. H. Orians, & K. Hyatt. 1976. Ecological implications of resource depression. *Amer. Nat.* 110:247–59.

LOTKA-VOLTERRA PRINCIPAL or MODEL [of predation]

Lotka, A. J. 1925. *Elements of Physical Biology.* Baltimore: Williams and Wilkins, chap. 8.

Volterra, V. 1931. Variations and fluctuations of the number of individuals in animal species living together. *Animal Ecology.* Edited by R. N. Chapman. New York: McGraw-Hill, pp. 409–48.

Pianka, E. R. 1976. Competition and niche theory. *Theoretical Ecology: Principles and Applications.* Oxford: Blackwell, pp. 114–41.

Wahgersky, P. J. 1978. Lotka-Volterra population models. *Annu. Rev. Ecol. Syst.* 9:189–218.

Cf. BIOLOGICAL EXPLOITATION THEORY; COM-

PETITIVE EXCLUSION; HASSELL-VARLEY MODEL; NICHOLSON-BAILEY MODEL; TIME-SPACE MODEL

LUDWIG HYPOTHESIS or EFFECT [ecology; population genetics]

= Ludwig's theorem

Ludwig, W. 1950. Zur Theorie der Konkurrenz. Die Annidation (Einnischung) als fünfter Evolutionsfaktor. *Neue Ergebnisse und Probleme Zoologie*. Edited by W. Herre. Leipzig: Geest & Portig, pp. 516–37.

Mayr, E. 1963. *Animal Species and Evolution*. Cambridge: Belknap Press of Harvard University Press, p. 245.

Soule, M. 1973. The epistasis cycle: a theory of marginal populations. *Annu. Rev. Ecol. Syst.* 4:165–87.

Cf. NICHE WIDTH-VARIATION HYPOTHESIS

LYON-RUSSELL HYPOTHESIS [embryogenesis]

Lyon, M. J. 1961. Gene action in the x chromosome of the mouse (*Mus musculus* L.). *Nature* 190:372–73.

______ . 1968. Chromosomal and subchromosomal inactivation. *Annu. Rev. Genet.* 2:31–48.

Russell, L. B. 1961. Genetics of mammalian sex chromosomes. *Science* 133:1795–1805.

Suzuki, D. T. 1974. Developmental genetics. *Concepts of development*. Edited by J. Lash & J. R. Whittaker. Stamford, Conn.: Sinauer Associates, pp. 349–79.

LYSENKOISM [genetics]

Lysenko, T. D. 1946. *Heredity and Its Variability*. New York: King's Crown.

Muller, H. J. 1948. The destruction of science in the USSR. *Sat. Rev. Lit.* Dec. 4, pp. 13–15, 63–65.

______ . 1949. It still isn't a science: a reply to George Bernard Shaw. *Sat. Rev. Lit.* April 16, pp. 11–12, 61.

Shaw, George Bernard. 1949. Behind the Lysenko controversy. *Sat. Rev. Lit.* April 16, p. 10.

Richens, R. H. 1950. Lysenko: the scientific issue. *New Biol.* 8:9–35.

Medvedev, Z. A. 1969. *The Rise and Fall of T. D. Lysenko*. Transl. by I. M. Lerner. New York: Columbia University Press.

Cf. LAMARCKIAN EVOLUTION; MICHURINISM

MACARTHUR MODEL *see* **BROKEN-STICK MODEL**

MACARTHUR-WILSON EQUILIBRIUM MODEL *see* **EQUILIBRIUM THEORY**

MCCOLLOUGH EFFECT [human color perception]
McCollough, C. 1965. Color adaptation of edge-detectors in the human visual system. *Science* 149:1115-16.
Shute, C. C. D. 1979. *The McCollough Effect.* Cambridge: Cambridge University Press.

MCLAREN EFFECT [ecology]
McLaren, I. A. 1963. Effects of temperature on growth of zooplankton and the adaptive value of vertical migration. *J. Fish. Res. Bd. Can.* 20:685–727.
______. 1974. Demographic strategy of vertical migration by a marine copepod. *Amer. Nat.* 108:91–102.
Beklemishev, C. W., A. A. Neyman, N. V. Parin, & H. J. Semina. 1972. Le biotope dans le milieu marin. *Marine Biology* 15:57–73.

MACROMOLECULAR CONCEPT [biochemistry]
Edsall, J. T. 1962. Proteins as macromolecules: an essay on the development of the macromolecular concept and some of its vicissitudes. *Arch. Biochem. Biophys.* (Suppl. 1):12–20.
Hess, E. L. 1970. Origins of molecular biology. *Science* 168:664–69.
Flory, P. J. 1973. Macromolecules vis-à-vis traditions of chemistry. *J. Chem. Educ.* 50:732–35.

MARGALEF'S PRINCIPLE [ecological succession]
Margalef, D. R. 1968. *Perspectives in Ecological Theory.* Chicago: University of Chicago Press.
Tansky, M. 1976. Structure, stability, and efficiency of ecosystem. *Prog. Theor. Biol.* 4:205–62.

MARGINAL POPULATION [genetics; evolution]
Carson, H. L. 1955. The genetic characteristics of marginal populations of *Drosophila. Cold Spr. Harb. Symp. Quant. Biol.* 20:276–87.
Soule, M. 1973. The epistasis cycle: a theory of marginal populations. *Annu. Rev. Ecol. Syst.* 4:165–87.

 MARGINAL VALUE THEOREM *see* **OPTIMAL FORAGING THEORY**

MARKER EFFECT [in genetic recombination]

Norkin, L. C. 1970. Marker-specific effects in genetic recombination. *J. Mol. Biol.* 51:633–55.

Stadler, D. R. 1973. The mechanism of intragenic recombination. *Annu. Rev. Genet.* 7:113–27.

MARSH'S LAWS [of evolution of brain size]

Marsh, O. C. 1874. Small size of the brain in Tertiary mammals. *Amer. J. Sci.*, ser. 3, 8:66–67.

Jerison, H. 1970. Brain evolution: new light on old principles. *Science* 170:1224–25.

Radinsky, L. 1978. Evolution of brain size in carnivores and ungulates. *Amer. Nat.* 112:815–31.

MASS FLOW CONCEPT [transport in plants]

Münch, E. 1930. *Die Stoffbewegungen in der Pflanze.* Jena: Fischer.

Zimmerman, M. H. 1974. Long distance transport. *Plant Physiol.* 54:472–79.

MASTER GENE *see* **MASTER-SLAVE HYPOTHESIS**

MASTER-SLAVE HYPOTHESIS [variation in DNA content]
= slave-repeats hypothesis

Callan, H. G. 1967. The organization of genetic units in chromosomes. *J. Cell Sci.* 2:1–7.

——— , & L. Loyd. 1960. Lampbrush chromosomes of crested newts *Triturus cristatus* (Laurenti). *Phil. Trans. Roy. Soc. Lond. B* 243:135–219.

Whitehouse, H. L. K. 1967. A cycloid model for the chromosome. *J. Cell Sci.* 2:9–22.

Thomas, C. A., Jr. 1970. The theory of the master gene. *The Neurosciences: Second Study Program.* Edited by F. O. Schmitt. New York: Rockefeller University Press, pp. 973–98.

Allen, S., & I. Gibson, 1972. Genome amplification and gene expression in the ciliate macronucleus. *Biochem. Genet.* 6:293–313.

Vorob'ev, V. I., S. N. Borchsenius, N. A. Belozerskaya, N. A. Merkulova, & I. S. Irlina. 1975. DNA replication in macronuclei of *Tetrahymena pyriformis* GL. *Exp. Cell Res.* 93:253–60.

Cf. NUCLEOSOME HYPOTHESIS; SUBNUCLEAR HYPOTHESIS

Kubitschek, H. E. 1964. Mutation without segregation. *Proc. Natl. Acad. Sci.* 52:1374–81.

Nasim, A., & C. Auerbach. 1967. The origin of complete and mosaic mutants from mutagenic treatment of single cells. *Mutat. Res.* 4:1–14.

Auerbach, C., & B. J. Kilbey. 1971. Mutation in eukaryotes. *Annu. Rev. Genet.* 5:163–218.

Cf. DUAL MECHANISM HYPOTHESIS; LETHAL-HIT HYPOTHESIS; REPAIR HYPOTHESIS

MATERIAL COMPENSATION, PRINCIPLE OF [embryogenesis; evolution]

= economy of growth energy = growth force

Rensch, B. 1959. *Evolution Above the Species Level.* New York: Columbia University Press, pp. 180*ff*.

Poulson, T. L. 1964. Animals in aquatic environments: animals in caves. *Handbook of Physiology.* Section 4. Edited by D. B. Dill. Washington, D.C.: American Physiological Society, pp. 749–71.

Barr, T. C., Jr. 1968. Cave ecology and the evolution of Troglobites. *Evol. Biol.* 2:35–102.

MATTHEW'S HYPOTHESIS [of peripheral populations; biogeography]

Matthew, W. D. 1915. Climate and evolution. *Ann. N.Y. Acad. Sci.* 24:171–318.

Ewan, J. 1944. The perennial southwestern Datura and the validity of Matthew's hypotheses in plant geography. *Proc. Cal. Acad. Sci.* IV 25:235–44.

Stebbins, G. L. 1950. *Variation and Evolution in Plants.* New York: Columbia University Press, pp. 533*ff*.

Cf. AGE AND AREA; GENE CENTERS

MAXIMUM HETEROZYGOSITY HYPOTHESIS [of overdominance; genetics]

Wallace, B. 1959. The role of heterozygosity in *Drosophila* populations. *Proc. 10th Int. Congr. Genet.* 1:408–19.

______. 1963. Further data on the overdominance of induced mutations. *Genetics* 48:633–51.

Murai, T. 1969. The genetic structure of natural populations of *Drosophila melanogaster.* VI. Further studies on the optimum heterozygosity hypothesis. *Genetics* 61:479–95.

Cf. OPTIMUM HETEROZYGOSITY HYPOTHESIS

MAXWELL'S DEMON [physiology; biophysics]

Maxwell, J. C. 1916. *The Theory of Heat*. 10th ed. London: Longmans, Green, p. 338.

Patlak, C. S. 1956. Contributions to the theory of active transport. *Bull. Math. Biophys.* 18:271-313.

Walker, I. 1976. Maxwell's demon in biological systems. *Acta Biotheor.* 25:103–10.

MAY'S PARADOX [ecology]

May, R. M. 1973. *Stability and Complexity in Model Ecosystems*. Princeton, N. J.: Princeton University Press, chap. 3*ff*.

Orians, G. H. 1975. Diversity, stability and maturity in natural ecosystems. *Unifying Concepts in Ecology*. 1st International Congress on Ecology, The Hague, Sept. 8–14, 1974. The Hague: Junk, pp. 139–50.

Cf. DIVERSITY-STABILITY THEORY

MECKEL-SERRES LAW [of recapitulation]

Russell, E. S. 1916. *Form and Function: A Contribution to the History of Animal Morphology*. London: Murray, p. 236.

Gould, S. J. 1977. *Ontogeny and Phylogeny*. Cambridge: Belknap Press of Harvard University Press, pp. 37*ff*.

Cf. BIOGENESIS

MEIOTIC COST [of sexual reproduction]

Williams, G. C. 1975. *Sex and Evolution*. Princeton, N.J.: Princeton University Press.

Barash, D. P. 1976. What does sex really cost? *Amer. Nat.* 110:894–97.

MEIOTIC DRIVE [genetics]

= segregation distortion

Sandler, L., & E. Novitski. 1957. Meiotic drive as an evolutionary force. *Amer. Nat.* 41:105–10.

______ , Y. Hiraizumi, & I. Sandler. 1959. Meiotic drive in natural populations of *Drosophila melanogaster*. I. The cytogenetic basis of segregation-distortion. *Genetics* 44:233–50.

Hartl, D. L. 1977. Applications of meiotic drive in animal breeding and population control. *Proceedings of the International Conference on Quantitative Genetics*. August 16–21, 1976. Edited by E. Pollak, O. Kempthorne, & T. B. Bailey. Ames: Iowa State University Press, pp. 63–88.

MENDELIAN CHROMOSOME THEORY *see* **CHROMOSOME THEORY**

MENDELIAN POPULATION [evolution]

Dobzhansky, T. 1950. Mendelian populations and their evolutions. *Amer. Nat.* 84:401–08.

______ . 1970. *Genetics of the Evolutionary Process.* New York: Columbia University Press, pp. 310*ff.*

Ehrlich, P. R. & R. W. Holm. 1962. Patterns and populations. *Science* 137:652–57.

MENDEL'S LAWS [of dominance, segregation, and independent assortment]

Mendel, G. 1866. Versuche über Pflanzenhybriden. *Verh. Naturforsch. ver. Brünn.* 4:3–47. (Reprinted in Engl. in *J. Roy. Hort. Soc.* 26:1–32, 1901–02.)

Bateson, W. 1909. *Mendel's Principles of Heredity.* Cambridge: Cambridge University Press.

Sturtevant, A. H. 1965. *A History of Genetics.* New York: Harper & Row.

Baxter, A. 1979. Mendel and meiosis. *J. Hist. Biol.* 12:137–73.

Olby, R. C. 1966. *Origins of Mendelism.* New York: Schocken, esp. chap. 6.

MERTENSIAN MIMICRY [in snakes; evolution]

Mertens, R. 1960. *The World of Amphibians and Reptiles.* London: Harrap.

______ . 1966. Das Problem der Mimikry bei Korallenschlangen. *Zool. Jahrb. Abt. Syst.* 84:541–76.

Wickler, W. 1968. *Mimicry in Plants and Animals.* New York: McGraw-Hill, chap. 12.

METABOLIC RATE LAW *see* **KLEIBER'S LAW**

METAGENESIS *see* **ALTERNATION OF GENERATIONS**

METAMORPHOSIS, THEORY OF [plant development]

Goethe, J. W. von. 1790. *Versuch die Metamorphose der Pflanzen zu erklären.* Gotha: Ettinger. Translated in Arber (1946).

Arber, A. 1946. *Goethe's Botany.* Waltham, Mass.: Chronica Botanica, vol. 10, no. 2, pp. 63–126.

______ . 1950. *The Natural Philosophy of Plant Form.* Cambridge: Cambridge University Press, pp. 41*ff.*

Guédès, M. 1969. La théorie de la métamorphose en morphologie végétale: des origines à Goethe et Batsch. *Rev. Hist. Sci. Applic.* 22:323–63.

______ . 1972. La théorie de la métamorphose en morphologie

végétale: A.-P. de Candolle et P.-J.-F. Turpin. *Rev. Hist. Sci. Applic.* 25:253–70.

Cf. CARPEL; PHYLLOTAXY

MEYRICK'S LAW *see* **DOLLO'S LAW**

MICHURINISM

Lysenko, T. D. 1948. Report and concluding remarks at the 1948 session of the Academy of Agricultural Sciences. (Engl. transl.: *Soviet Biology*, London: Birch.)

Huxley, J. 1949. *Heredity East and West: Lysenko and World Science*. New York: Henry Schuman, p. 16.

Cf. LAMARCKIAN EVOLUTION; LYSENKOISM

MICRONUTRIENT HYPOTHESIS [of primate feeding behavior]

Hamilton, W. J., III, & C. D. Busse. 1978. Primate carnivory and its significance to human diets. *Bioscience* 28:761–66.

MIGRATION HYPOTHESIS [of bipolar distribution of animals]

Ortmann, A. E. 1899. On new facts lately presented in opposition to the hypothesis of bipolarity of marine faunas. *Amer. Nat.* 33:583–91.

Allee, W. C., & K. P. Schmidt. 1951. *Ecological Animal Geography*. 2d ed., rev. by R. Hesse. New York: Wiley, pp. 333*ff*.

MIGRATIONAL SELECTION [behavior; evolution]

= migrational polymorphism

Parsons, P. A. 1963. Migration as a factor in natural selection. *Genetics* 33:184–206.

Doyle, R. W. 1976. Analysis of habitat loyalty and habitat preference in the settlement behavior of planktonic marine larvae. *Amer. Nat.* 110:719–30.

MILIEU INTÉRIEUR *see* **BERNARD'S PRINCIPLE**

MIMETIC POLYMORPHISM [evolution]

Goldschmidt, R. B. 1945. Mimetic polymorphism, a controversial chapter of Darwinism. *Q. Rev. Biol.* 20:147–64; 205–30.

Carpenter, G. D. H. 1946. Mimetic polymorphism. *Nature* 158:277–79.

Ford, E. B. 1953. The genetics of polymorphism in the lepidoptera. *Adv. Genet.* 5:43–87.
Cf. MIMICRY

MIMICRY [evolution]

Punnett, R. C. 1915. *Mimicry in Butterflies.* Cambridge: Cambridge University Press.

Pijl, L. van der, & C. H. Dodson. 1967. *Orchid Flowers, Their Pollination and Evolution.* Coral Gables, Fla.: University of Miami Press, chap. 11.

Wickler, W. 1968. *Mimicry in Plants and Animals.* New York: McGraw-Hill.

Brower, L. P., & J. van Z. Brower. 1972. Parallelism, convergence, divergence, and the new concept of advergence in the evolution of mimicry. *Trans. Conn. Acad. Arts Sci.* 44:57–67.

Barlow, B. A., & D. Wiens. 1977. Host-parasite resemblance in Australian mistletoes: the case for cryptic mimicry. *Evolution* 31:69–84.

Wiens, D. 1978. Mimicry in plants. *Evol. Biol.* 11:365–403.

Cf. BATESIAN MIMICRY; MERTENSIAN MIMICRY; MÜLLERIAN MIMICRY; WALLACE'S RULES OF MIMICRY

MINIMAL AREAS, LAW OF [cellular arrangement]

Lamarle, E. 1864–67. Sur la stabilité des systèms liquides en lames minces. *Mém. Acad. Roy. Sci. Lett. Belg.* 35, pt. 2; 36, pt. 6.

Thompson D'A. W. 1942. *On Growth and Form.* 2d ed. Cambridge: Cambridge University Press, p. 486.

MINIMUM, LAW OF [ecology]

Liebig, J. 1846. *Animal Chemistry, or Chemistry in its Application to Physiology and Pathology.* 3d ed. London: Taylor & Walton.

Blackman, F. F. 1905. Optima and limiting factors. *Ann. Bot.* 19:281–95.

Taylor, W. P. 1934. Significance of extreme or intermittent conditions in distribution of species and management of natural resources, with a restatement of Liebig's law of minimum. *Ecology* 15:374–79.

Rübel, E. 1935. The replaceability of ecological factors and the law of minimum. *Ecology* 16:336–41.

Pianka, E. R. 1978. *Evolutionary Ecology.* 2d ed. New York: Harper & Row, chap. 1.

Cf. TOLERANCE, LAW OF

MINIMUM EVOLUTION, PRINCIPLE OF

Edwards, A. W. F., & L. L. Cavalli-Sforza. 1963. The reconstruction of evolution. (Abstract of paper.) *Ann. Hum. Genet.* 27:105, and *Heredity* 18:553.

Cavalli-Sforza, L. L., & A. W. F. Edwards. 1967. Phylogenetic analysis: models and estimation procedures. *Evolution* 21:550–70.

Sneath, P. H., & R. R. Sokal, 1973. *Numerical Taxonomy.* San Francisco: Freeman, chap. 6.

Cf. PARSIMONY

MINIMUM GENETIC LOAD, PRINCIPLE OF

Kimura, M. 1960. Optimum mutation rate and degree of dominance as determined by the principle of minimum genetic load. *J. Genet.* 57:21–34.

MINIMUM WORK, PRINCIPLE OF [physiology]

= minimal work, principle of

Murray, C. D. 1926. The physiological principle of minimum work. I. The vascular system and the cost of blood volume. *Proc. Natl. Acad. Sci.* 12:207–14.

______ . 1926. The physiological principle of minimum work applied to the angle of branching of arteries. *J. Gen. Physiol.* 9:835–41.

Thompson, D'A. W. 1942. *On Growth and Form.* 2d ed. Cambridge: Cambridge University Press, pp. 949–50.

MISTAKE POLLINATION [ecology]

Baker, H. G. 1976. "Mistake" pollination as a reproductive system with special reference to the Caricaceae. *Tropical Trees.* Edited by J. Burley & B. T. Styles. New York: Academic Press, pp. 161–69.

MITCHELL EFFECT *see* **GENE CONVERSION**

MODERN SYNTHESIS *see* **SYNTHETIC THEORY** [evolution]

MODIFIED-ENZYME THEORY [of antibody formation]

Burnet, F. M., & F. Fenner. 1949. *The Production of Antibodies.* 2d ed. Melbourne: Macmillan, chap. 8.

Jerne, N. K. 1955. The natural-selection theory of antibody formation. *Proc. Natl. Acad. Sci.* 41:849–57.

MODIFIED PELTATION THEORY [of the tubular leaf of *Nepenthes*]

Franck, D. H. 1976. The morphological interpretation of epiascidate leaves. *Biol. Rev.* 42:345–88.

MODIFIER GENE [genetics]

Karlin S., & J. McGregor. 1974. Towards a theory of the evolution of modifier genes. *Theor. Pop. Biol.* 5:59–103.

Feldman, M. W., & J. Krakauer. 1976. Genetic modification and modifier polymorphisms. *Population Genetics and Ecology*. Edited by S. Karlin & E. Nevo. New York: Academic Press, pp. 547–83.

"MOLE" HYPOTHESIS [cicada predation]

Lloyd, M., & H. S. Dybas. 1966. The periodical cicada problem. I. Population ecology. II. Evolution. *Evolution* 20:133–49, 466–505.

MOLECULAR CLOCK [primate phylogeny]

Zuckerkandl, E., & L. Pauling. 1965. Evolutionary divergence and convergence in proteins. *Evolving Genes and Proteins*. Edited by V. Bryson & H. J. Vogel. New York: Academic Press, pp. 97–166.

Read, D. W. 1975. Primate phylogeny, neutral mutations, and "molecular clocks." *Syst. Zool.* 24:209–21.

MOLECULAR EVOLUTIONARY CLOCK [time-related changes in divergent evolution]

= evolutionary clock

Sarich. V. M. 1970. Primate systematics with special reference to old world monkeys: a protein perspective. *Old World Monkeys: Evolution, Systematics, and Behavior*. Edited by J. R. Napier & P. H. Napier. New York: Academic Press, pp. 175–226.

Jukes, T. H., & R. Holmquist. Evolutionary clock: nonconstancy of rate in different species. *Science* 177:530–32.

Fitch, W. M. 1976. Molecular evolutionary clocks. *Molecular Evolution*. Edited by F. J. Ayala. Sunderland, Mass.: Sinauer Associates, pp. 160–78.

———, & G. H. Langley. 1978. Protein evolution and the molecular clock. *Evolution of Protein Molecules*. Edited by H. Matsubara & T. Yamanaka. Tokyo: Japan Scientific Societies, pp. 45–60.

MOLECULAR MIMICRY [parasitology]

= antigenic similarity

Dineen, J. K. 1963. Antigenic relationships between host and parasite. *Nature* 197:471–72.

Damian, R. T. 1964. Molecular mimicry: antigen sharing by parasite and host and its consequences. *Amer. Nat.* 98:129–49.

Read, C. P. 1972. *Animal Parasitism*. Englewood Cliffs, N.J.: Prentice-Hall, pp. 165–67.

MONOCLIMAX *see* **CLIMAX**

MONOGENIC HETEROSIS *see* **CONDITIONAL HETEROSIS**

MONOPHYLESIS *see* **MONOPHYLY PRINCIPLE**

MONOPHYLY PRINCIPLE [evolution]

Hennig, W. 1966. *Phylogenetic Systematics*. Urbana: University of Illinois Press, chap. 3.

Tuomikoski, R. 1967. Notes on some principles of phylogenetic systematics. *Ann. Entomol. Fenn.* 33:137–47.

Nelson, G. J. 1971. Paraphyly and polyphyly: redefinitions. *Syst. Zool.* 20:471–72.

Ashlock, P. D. 1971. Monophyly and associated terms. *Syst. Zool.* 20:63–69.

______. 1972. Monophyly again. *Syst. Zool.* 21:430–38.

Colless, D. H. 1972. A note on Ashlock's definition of "monophyly." *Syst. Zool.* 21:126–28.

Cf. POLYPHYLETIC THEORY

MONOTONY PRINCIPLE [repetition in bird songs]

Hartshorne, C. 1956. The monotony-threshold in singing birds. *Auk* 73:176–92.

Cf. TWO-FACTOR THEORY OF INHIBITION

MORES [ecological species]

Shelford, V. E. 1912. Ecological succession. V. Aspects of physiological classification. *Biol. Bull.* 23:331–70.

Cf. ECOSPECIES

MORGAN'S CANON [animal behavior]

Morgan, C. L. 1894. *An Introduction to Comparative Psychology*. London: Scott, p. 53.

Allee, W. C., A. E. Emerson, O. Park, T. Park, & K. P. Schmidt. 1949. *Principles of Animal Ecology*. Philadelphia: Saunders, p. 5.

Cf. OCCAM'S RAZOR

MORGAN'S LAW [of a linear gene disposition in chromosomes]

Morgan, T. H. 1919. *The Physical Basis of Heredity*. Philadelphia: Lippincott, chap. 9.

Roemer, T. 1936. Bedeutung des Gesetzes der Parallelvariation für die Pflanzenzüchtung. *Nova Acta Leopoldina* 4:351–65.

MORPHOLOGICAL CORRESPONDENCE [morphology; evolution]

Woodger, J. H. 1945. On biological transformations. *Essays on Growth and Form.* Edited by W. E. LeGros Clark & P. B. Medawar. Oxford: Oxford University Press, pp. 95–125.

Withers, R. F. J. 1964. Morphological correspondence and the concept of homology. *Form and Strategy in Science.* Edited by J. R. Gregg & F. T. C. Harris. Dordrecht, Netherlands: Reidel, pp. 378–94.

Cf. HOMOLOGY

MORPHOLOGICAL SPECIES CONCEPT *see* **SPECIES CONCEPTS**

MOSAIC EVOLUTION

= Watson's rule = mosaic inheritance theory

Roux, W. 1883. *Über die Bedeutung der Kerntheilungsfiguren.* Leipzig: Engelmann.

Watson, D. M. S. 1918. On *Seymouria,* the most primitive known reptile. *Proc. Zool. Soc. Lond.* 1918:267–301.

Wilson, E. B. 1925. *The Cell in Development and Heredity.* New York: Macmillan, pp. 1042–50.

de Beer, G. R. 1954. Archaeopteryx and evolution. *Advmt. Sci.* 11:160–70.

Cracraft, J. 1970. Mandible of *Archaeopterix* provides an example of mosaic evolution. *Nature* 226:1268.

MOSAIC THEORY OF REGENERATION [ecology-succession]

Aubreville, A. 1938. La foret coloniale: les forets de l'Afrique occidentale francaise. *Ann. Acad. Sci. Colon.* 9:1–245.

Connell, J. H., & R. O. Slayter. 1977. Mechanisms of succession in natural communities and their role in community stability and organization. *Amer. Nat.* 111:1119–44.

Cf. CYCLIC SUCCESSION; SUCCESSION

MOTIVATION [behavior]

Weiner, B. 1972. *Theories of Motivation.* Chicago: Rand McNally.

Cf. DRIVE

MOTIVATION-STRUCTURAL (MS) RULES [animal communication]

Morton, E. S. 1977. On the occurrence and significance of motivation-structural rules in some bird and mammal sounds. *Amer. Nat.* 111:855–69.

 MÜLLERIAN MIMICRY [evolution]

Müller, F. 1878. Über die Vortheide der Mimikry bei Schmellerlingen. *Zool. Anz.* 1:54–55.

______ . 1879. *Ithuna* and *Thyridia;* a remarkable case of mimicry in butterflies. *Proc. Entomol. Soc. Lond.* 1879: xx-xxix.

Sheppard, P. M. 1959. The evolution of mimicry: a problem in ecology and genetics. *Cold Spr. Harb. Symp. Quant. Biol.* 24:131–40.

Brower, L. P., J. V. Z. Brower, & C. T. Collins. 1963. Experimental studies of mimicry. 7. Relative palatability and Müllerian mimicry among neotropical butterflies of the subfamily Heliconiinae. *Zoologica* (N.Y.) 48:65–83.

Turner, J. R. G. 1971. Studies of Müllerian mimicry and its evolution in Burnet moths and heliconid butterflies. *Ecological Genetics and Evolution.* Edited by R. Creed. Oxford: Blackwell, pp. 224–60.

Huheey, J. E. 1976. Studies in warming coloration and mimicry. VII. Evolutionary consequences of a Batesian-Müllerian spectrum: a model for Müllerian mimicry. *Evolution* 30:86–93.

Cf. MIMICRY

MÜLLER'S LAW [neuron activity]

Müller, J. 1835. *Handbuch der Physiologie des Menehen für Vorlesungen.* 2d ed. vol. I, book 3, sect. 4; Coblenz.

Riese, W., & G.E. Arrington, Jr. 1963. The history of Johannes Müller's doctrine of the specific energies of the senses: original and later versions. *Bull. Hist. Med.* 37:179–83.

Horridge, G. A. 1977. Mechanistic teleology and explanation in neuroethology. *Bioscience* 27:725–32.

MÜLLER'S LAW [spine arrangement in Radiolaria]

Müller, J. 1826. *Zur vergleichenden Physiologie des Gesichtssinnes des Menschen und der Thiere.* Leipzig: Cnobloch, pp. 44–45.

______ . 1858. Über die Thalassicollen, Polycistinen und Acanthrometren des Mittelmeeres. *Abh. Akad. Wiss.*, pp. 1–62.

Thompson, D'A. W. 1942. *On Growth and Form.* 2d ed. Cambridge: Cambridge University Press, p. 725.

Woodward, W. E. 1975. Hermann Lotze's critique of Johannes Müller's doctrine of specific sense energies. *Med. Hist.* 19:147–57.

MÜLLER'S PHYTOALEXIN THEORY *see* **PHYTOALEXIN THEORY**

MULTIAXIAL FLOWER THEOREM [morphology]
= pseudanthium hypotheses

Wettstein, R. von. 1907. Die Entwicklung der Blüte der angiospermen Pflanzen aus der jenigen der Angiospermen. *Das Wissen für alle*, Jahrg. 1907, no. 45, pp. 705–08.

______ . 1907. *Handbuch der systematischen Botanik.* II. Leipzig & Vienna: Deuticke.

Meeuse, A. D. J. 1972. Sixty-five years of theories of the multiaxial flower. *Acta Biotheor.* 21:167–202.

MULTICELLULAR THEORY [of nerve fiber development]

Schwann T. 1839. *Mikroskopische Untersuchungen über die Uebereinstimmung in der Struktur und dem Wachsthum der Thiere und Pflanzen.* Berlin: Sanders.

Billings, S. M. 1971. Concepts of nerve fiber development, 1839–1930. *J. Hist. Biol.* 4:275–305.

Cf. NEURON THEORY; OUTGROWTH THEORY; PROTOPLASMIC BRIDGE THEORY

MULTI-FACTORIAL THEORY *see* **MULTIPLE FACTOR HYPOTHESIS**

MULTIPLE-ALLELE, SINGLE-LOCUS HYPOTHESIS [genetics]

Whiting, P. W. 1939. Sex determination and reproductive economy in *Habrobracon. Genetics* 24:110–11.

Crozier, R. H. 1971. Heterozygosity and sex determination in haplo-diploidy. *Amer. Nat.* 105:399–412.

Cf. GENIC-BALANCE HYPOTHESIS; MULTIPLE-LOCUS HYPOTHESIS

MULTIPLE FACTOR HYPOTHESIS [genetics]
= polymeric inheritance = multi-factorial theory

Nilsson-Ehle, H. 1909. Kreuzungsuntersuchungen an Hafer und Weizen. *Lunds Univ. Arssk.* N. F. Afd. 2 5:1–122.

East, E. M. 1910. A Mendelian interpretation of variation that is apparently continuous. *Amer. Nat.* 44:65–82.

Carlson, E. A. 1966. *The Gene: A Critical History.* Philadelphia: Saunders, chap. 4.

Norton, B. 1975. Metaphysics and population genetics: Karl Pearson and the background to Fisher's multi-factorial theory of inheritance. *Ann. Sci.* 32:537–53.

Cf. POLYGENE

MULTIPLE FOLIAR HELICES THEORY [plant morphology]

Plantefol, L. 1946–47. Fondements d'un theorie phyllotaxique nouvelle. I-IV. *Ann. Sci. Nat. Bot.*, ser. II, 7:153–229; 8:1–71.

Cutter, E. G. 1959. On a theory of phyllotaxis and histogenesis. *Biol. Rev.* 34:243–63.

MULTIPLE GENE EFFECTS *see* **PLEITROPHY**

MULTIPLE-LOCUS HYPOTHESIS [genetics]

Snell, G. D. 1935. The determination of sex in *Habrobracon*. *Proc. Natl. Acad. Sci.* 21:446–53.

Crozier, R. H. 1971. Heterozygosity and sex determination in haplo-diploidy. *Amer. Nat.* 105:399–412.

Lewontin, R. C. 1973. Population genetics. *Annu. Rev. Genet.* 7:1–17.

Cf. GENIC-BALANCE HYPOTHESIS; MULTIPLE-ALLELE, SINGLE-LOCUS HYPOTHESIS

MULTIPLE SISTER-GROUP RULE [biogeography]

Brundin, L. 1966. Transantarctic relationships and their significance, as evidenced by the chironomid midges, with a monograph of the subfamily Podonominae, Aphrotaeninae and the austral Heptagiae. *K. Sven. Vetenskapsakad. Handl.* 11:1–472.

Hennig, W. 1966. The Diptera fauna of New Zealand as a problem in systematics and zoogeography. *Pac. Insects Monogr.* 9:1–81.

Ashlock, P. D. 1974. The uses of cladistics. *Annu. Rev. Ecol. Syst.* 5:81–99.

MURDER [of one chromosome by another] *see* **OUTLAW GENES**

MURPHY'S LAW [of beak size in birds]

Murphy, R. C. 1938. The need of insular exploration as illustrated by birds. *Science* 88:533–39.

Lack, D. 1947. *Darwin's Finches.* Cambridge: Cambridge University Press, p. 80.

Grant, P. R. 1965. The adaptive significance of some size trends in island birds. *Evolution* 19:355–67.

Power, D. M. 1971. Evolution of the house finch on Santa Cruz island, California. *Can. J. Zool.* 49:675–84.

MUSEUM HYPOTHESIS [biogeography; evolution]

Stebbins, G. L. 1974. *Flowering Plants: Evolution Above the Species Level.* Cambridge: Harvard University Press, p. 166.

MUTATION-EQUILIBRIUM THEORY [of molecular evolution]

King, J. L., & T. H. Ohta. 1975. Polyallelic mutational equilibria. *Genetics* 79:681–91.

MUTATION THEORY [evolution]

de Vries, H. 1901–03. *Die Mutationstheorie*. 2 vols. Leipzig: Veit.

Castle, W. E. 1905. The mutation theory of organic evolution from the standpoint of animal breeding. *Science* 21:521–25.

Muller, H. J. 1958. The mutation theory re-examined. *Proc. 10th Int. Congr. Genet.*, pp. 306–17.

Allen, G. E. 1969. Hugo deVries and the reception of the "Mutation Theory." *J. Hist. Biol.* 2:55–87.

Dobzhansky, T. 1970. *Genetics of the Evolutionary Process*. New York: Columbia University Press, chap. 2.

Cf. SALTATION

MUTUAL EXCLUSION EFFECT [elimination of one bacteriophage by another]

Delbrück, M. 1945. Interference between bacterial viruses. III. The mutual exclusion effect and the depressor effect. *J. Bacteriol.* 50:151–70.

MUTUAL FACILITATION [population genetics]

Beardmore, J. A. 1963. Mutual facilitation and the fitness of polymorphic populations. *Amer. Nat.* 97:69–74.

Soule, M. 1973. The epistasis cycle: a theory of marginal populations. *Annu. Rev. Ecol. Syst.* 4:165–87.

MUTUALISM [ecology; evolution]

Van Beneden, J. P. 1876. *Animal Parasites and Messmates*. New York: Appleton, chap. 4.

de Bary, A. 1887. *Comparative Morphology and Biology of the Fungi, Mycetozoa and Bacteria*. Oxford: Oxford University Press, p. 369.

Wheeler, W. M. 1923. *Social Life Among the Insects*. New York: Harcourt, Brace, pp. 195*ff*.

Allee, W. C., A. E. Emerson, O. Park, T. Park, & K. P. Schmidt. 1949. *Principles of Animal Ecology*. Philadelphia: Saunders, chaps. 17, 35.

Rubtzov, I. A. 1971. Protective and mutualistic relationships between the insect and its parasites. (In Russian.) *Zh. Obshch. Biol.* 32:193–203.

Gilbert, L. E. 1975. Ecological consequences of a coevolved mutualism between butterflies and plants. *Coevolution of*

Animals and Plants. Edited by L. E. Gilbert & P. H. Raven. Austin: University of Texas Press, pp. 210–39.

Goh, B. S. 1979. Stability in models of mutualism. *Amer. Nat.* 113:261–75.

Cf. BIOTROPHY; COEVOLUTION; SYMBIOSIS

M-V LINKAGE [genetics]

Grant, V. 1967. Linkage between morphology and viability in plant species. *Amer. Nat.* 101:125–39.

______. 1971. *Plant Speciation*. New York: Columbia University Press, pp. 65, 175.

[N]

NAKED SEED [plant morphology]

Lorch, J. W. 1959. Gleanings on the naked seed controversy. *Centaurus* 6:122–28.

NASTIES or NASTIC RESPONSE [plant physiology]

Weevers, T. H. 1949. *Fifty Years of Plant Physiology*. Amsterdam: Scheltema & Holkema; Waltham, Mass.: Chronica Botanica, pp. 231–60.

NATURAL AFFINITY CONCEPT [taxonomy]

Lindley, J. 1836. *A Natural System of Botany*. 2d ed. London: Longman.

Gilmour, J. S. L. 1951. The development of taxonomic theory since 1851. *Nature* 168:400–02.

NATURAL BALANCE *see* **BALANCE OF NATURE**

NATURAL CONTROL THEORIES [ecology; population control]

Thompson, W. R. 1929. On natural control. *Parasitology* 21:269–81.

Thompson, W. R. 1939. Biological control and the theories of the interactions of populations. *Parasitology* 31:299–388.

______. 1956. The fundamental theory of natural and biological control. *Annu. Rev. Entomol.* 1:379–402.

Milne, A. 1957. Theories of natural control of insect populations. *Cold Spr. Harb. Symp. Quant. Biol.* 22:253–71.

Cf. COMPETITIVE EXCLUSION; LOTKA-NICHOLSON THEORY; LOTKA-VOLTERRA PRINCIPLE; VOLTERRA's LAW

NATURAL SELECTION [evolution]

Darwin, C. 1859. *On the Origin of Species by Means of Natural Selection*. London: Murray.

Pearl, R. 1917. The selection problem. *Amer. Nat.* 51:65–91.

Fisher, R. A. 1930. *The Genetical Theory of Natural Selection*. Oxford: Clarendon Press.

——— . 1954. Retrospect of the criticisms of the theory of natural selection. *Evolution as a Process*. Edited by J. Huxley, A. C. Hardy, & E. B. Ford. London: Allen & Unwin, pp. 84–98.

Lerner, I. M. 1959. The concept of natural selection: a centennial view. *Proc. Amer. Philos. Soc.* 103:173–82.

Mason, H. L., & J. H. Langenheim. 1961. Natural selection as an ecological concept. *Ecology* 42:158–65.

Dickinson, A. 1964. *Charles Darwin and Natural Selection*. New York: Watts.

Li, C. C. 1967. Fundamental theorem of natural selection. *Nature* 214:505–06.

Beddall, B. G. 1968. Wallace, Darwin, and the theory of natural selection. *J. Hist. Biol.* 1:261–323.

Barker, A. D. 1969. An approach to the theory of natural selection. *Philosophy* 44:271–90.

Vorzimmer, P.J. 1969. Darwin, Malthus, and the theory of natural selection. *J. Hist. Ideas* 30:527–42.

Van Valen, L. 1976. Energy and evolution. *Evol. Theory* 1:179–229.

NATURAL SELECTION, LAW OF *see* **FISHER'S FUNDAMENTAL THEOREM; THIRD LAW OF NATURAL SELECTION**

NATURAL SELECTION THEORY [of antibody formation]

Jerne, N.K. 1955. The natural-selection theory of antibody formation. *Proc. Natl. Acad. Sci.* 41:849–57.

——— . 1966. The natural selection theory of antibody formation: ten years later. *Phage and the Origins of Molecular Biology*. Edited by J. Cairns, G. S. Stent, & J. D. Watson. Cold Spring Harbor, N.Y.: Cold Spring Harbor Laboratory of Quantitative Biology, pp. 301–12.

Edelman, G.M. 1974. The problem of molecular recognition by a selective system. *Studies in the Philosophy of Biology*. Edited by F. J. Ayala & T. Dobzhansky. Berkeley: University of California Press, pp. 45–56.

Cf. SELECTION HYPOTHESIS

NAVASHINIAN FUSION [of chromosomes]

Navashin, M. 1933. The dislocation hypothesis of evolution of chromosome numbers. *Z. Indukt. Abstammungs-Vererbungsl.* 63:224–31.

Small, E. 1971. The evolution of reproductive isolation in *Clarkia*, section *Myxocarpa*. *Evolution* 25:330–46.

NEIGHBOR EFFECT [evolution of behavior]

Eshel, I. 1972. On the neighbor effect and the evolution of altruistic traits. *Theor. Pop. Biol.* 3:258–77.

Cf. ALTRUISTIC TRAITS

NEO-BERGMANNIAN RULE [faunistic distribution]

James, F.C. 1970. Geographic size variation in birds and its relationship to climate. *Ecology* 51:365–90.

Cf. BERGMANN'S RULE

NEO-DARWINIAN EVOLUTION

= SYNTHETIC THEORY = adaptive neutrality

Mayr, E. 1963. *Animal Species and Evolution.* Cambridge: Belknap Press of Harvard University Press, pp. 3*ff*.

Waddington, C.H. 1968. The paradigm for the evolutionary process. *Population Biology and Evolution.* Edited by R. C. Lewontin. Syracuse, N.Y.: Syracuse University Press, pp. 37–45.

Maynard Smith, J. 1969. The status of neo-Darwinism. *Towards a Theoretical Biology.* Vol. II: *Sketches.* Edited by C. H. Waddington. Edinburgh: Edinburgh University Press, pp. 82–89.

Richmond, R.C. 1970. Non-Darwinian evolution: a critique. *Nature* 225:1025–28.

King, J.L. 1972. The role of mutation in evolution. *Proc. 6th Berkeley Symp. Math. Stat. Probab.* 5:69–100.

Cf. GENETIC DRIFT; NON-DARWINIAN EVOLUTION; NEUTRAL THEORY

NEOTENY [morphogenesis; evolution]

de Beer, G.R. 1958. *Embryos and Ancestors.* 3d ed. Oxford: Clarendon Press, chap. 8.

Stebbins, G.L. 1974. *Flowering Plants: Evolution Above the Species Level.* Cambridge: Harvard University Press, p. 116.

Takhtajan, A. 1976. Neoteny and the origin of flowering plants. *Origin and Early Evolution of Angiosperms.* Edited by C. B. Beck. New York: Columbia University Press, pp. 207–19.

Gould, S.J. 1977. *Ontongeny and Phylogeny.* Cambridge:

Harvard University Press, pp. 341*ff*.
Cf. DARWIN'S ABOMINABLE MYSTERY

NERVE-NET THEORY [of intercellular communication]
see **RETICULAR THEORY**

NEUGLIEDERUNG CONCEPT [of vertebral body boundaries]

Remak, R. 1855. *Untersuchungen über die Entwicklung der Wirbelthiere*. Berlin: Reimer.

Verbout, A. J. 1976. A critical review of the "Neugliederung" concept in relation to the development of the vertebral column. *Acta Biotheor*. 25:219–58.

NEURON THEORY [of intercellular communication]
= contact theory = neurone theory

Waldeyer, H. W. G. 1891. Über einige neure Forschungen im Gebiete den Anatomie des Centralnervensystems. *Deut. Med. Woch*. 17:1213–18, 44–46, 67–69, 87–89, 1331–32, 52–56.

Golgi, C. 1906. The neuron doctrine—theory and facts. *Nobel Lectures: Physiology or Medicine, 1901–1921*. Amsterdam: Elsevier (1967), pp. 189–217.

Cajal, S. Ramon y. 1954. *Neuron Theory or Reticular Theory?* Transl. by M. O. Purkiss & C. A. Fox. Madrid: Consejo Superior de Investigationes Cientificas Instituto "Ramon y Cajal."

French, R. D. 1970. Some concepts of nerve structure and function in Britain, 1875–1885: background to Sir Charles Sherrington and the synapse concept. *Med. Hist*. 14:154–65.

Grundfest, H. 1975. History of the synapse as a morphological and functional structure. *Golgi Centennial Symposium*. Edited by M. Santini. New York: Raven, pp. 39–50.

Cf. INTEGRATIVE ACTION

NEUTRAL THEORY [of genetic fitness or variation in evolution]
= neo-classical theory = neutral allele hypothesis = neutral mutation hypothesis = neutral-mutation-random-drift theory = neutral variation hypothesis = neutralist view = neutrality theory

Kimura, M. 1968. Evolutionary rate at the molecular level. *Nature* 217:624–26.

———. 1979. The neutral theory of molecular evolution. *Sci. Amer*. 241 (5): 98–126.

King, J. L., & T. H. Jukes. 1969. Non-Darwinian evolution. *Science* 164:788–98.

Crow, J.F. 1972. The dilemma of nearly neutral mutations: how important are they for evolution and human welfare? *J. Hered.* 63:306–16.

Ayala, F.J. 1974. Biological evolution: natural selection or random walk? *Amer. Sci.* 62:692–701.

Aspinwall, N. 1974. Genetic analysis of North American populations of the pink salmon, *Oncorhynchus gorbuscha*, possible evidence for the neutral mutation-random drift hypothesis. *Evolution* 28:295–305.

Lewontin, R.C. 1974. *The Genetic Basis of Evolutionary Change*. New York: Columbia University Press, pp. 197*ff*.

Redfield, J.A. 1974. Genetics and selection at the Ng locus in blue grouse (*Dendragapus obscurus*). *Heredity* 33:69–78.

Nei, M. 1975. *Molecular Population Genetics and Evolution*. New York: American Elsevier, pp. 5*ff*.

Zangerl, A R., S.T.A. Pickett, & F.A. Bayzaz. 1977. Some hypotheses on variation in plant populations and an experimental approach. *Biologist* 59:113–22.

Maynard Smith, J. 1978. Optimization theory in evolution. *Annu. Rev. Ecol. Syst.* 9:31–56.

Cf. GENETIC DRIFT; NON-ADAPTIVE EVOLUTION; NON-DARWINIAN EVOLUTION; RANDOM WALK; SELECTIONIST VIEW; STEPPING-STONE MODEL

NEWTON'S LAW [of cooling]

Newton, I. 1744. *Scala Graduum Caloris et Frigoris*. Lausanne.

Kleiber, M. 1961. *The Fire of Life*. New York: Wiley, chap. 8.

NICHE [ecology]

Grinnell, J. 1904. The origin and distribution of the chestnut-backed chickadee. *Auk* 21:364–82.

______ . 1917. The niche relationships of the California thrasher. *Auk* 34:427–33.

Johnson, R.H. 1910. Determinate evolution in the color pattern of the lady-beetles. *Carnegie Inst. Wash. Publ.* 122.

Elton, C. 1927. *Animal Ecology*. London: Sidgwick & Jackson, p. 63.

Geisel, T. S. (Dr. Seuss). 1955. *On Beyond Zebra*. New York: Random House, "Nuh."

Hutchinson, G.E. 1965. *The Ecological Theatre and the Evolutionary Play*. New Haven, Conn.: Yale University Press, pp. 26*ff*.

______ . 1978. *An Introduction to Population Ecology*. New Haven, Conn.: Yale University Press, pp. 152*ff*.

Levins, R. 1968. *Evolution in Changing Environments*. Princeton, N.J.: Princeton University Press, chap. 3.

MacArthur, R.H. 1968. The theory of the niche. *Population Biology and Evolution*. Edited by R. Lewontin. Syracuse, N.Y.: Syracuse University Press, pp. 159–70.

Levin, S. A. 1970. Community equilibria and stability, and an extension of the competitive exclusion principle. *Amer. Nat.* 104:413–23.

Whittaker, R.H. 1975. *Communities and Ecosystems*. 2d ed. New York: Macmillan, pp. 77–87.

———, S.A. Levin, & R.B. Root. 1973. Niche, habitat, and ecotope. *Amer. Nat.* 107:321–38.

Pianka, E.R. 1976. Competition and niche theory. *Theoretical Ecology: Principles and Application*. Edited by R. M. May. Oxford: Blackwell, chap. 7.

Cf. COMPETITIVE EXCLUSION

NICHE EXPANSION [ecology]

= ecological release

Wilson, E.O. 1961. The nature of the taxon cycle in the Melanesian ant fauna. *Amer. Nat.* 95:169–93.

Lister, B.C. 1976. The nature of niche expansion in West Indian *Anolis* lizards. I. Ecological consequences of reduced competition. II. Evolutionary components. *Evolution* 30:659–76, 677–92.

NICHE HYPERSPACE [ecology]

Hutchinson, G.E. 1957. Concluding remarks. *Cold Spr. Harb. Symp. Quant. Biol.* 22:415–27.

Whittaker, R. H. 1969. Evolution of diversity in plant communities. *Brookh. Symp. Biol.* 22:178–96.

———. 1972. Evolution and measurement of species diversity. *Taxon* 21:213–51.

NICHE OVERLAP HYPOTHESIS [ecology]

MacArthur, R.H., & R. Levins. 1967. The limiting similarity, convergence, and divergence of coexisting species. *Amer. Nat.* 101:377–85.

Pianka, E. R. 1972. r and K selection or b and d selection? *Amer. Nat.* 106:581–89.

May, R. M. 1974. On the theory of niche overlap. *Theor. Pop. Biol.* 5:279–332.

Abrams, P. A. 1977. Density-independent mortality and interspecific competition: a test of Pianka's niche overlap hypothesis. *Amer. Nat.* 111:539–52.

Cf. BROKEN-STICK MODEL

NICHE SHIFT [ecology]

Lack, D., & H. N. Southern. 1949. Birds on Tenerife. *Ibis* 91:607–26.

Werner, E. E., & D. J. Hall. 1976. Niche shifts in sunfishes: experimental evidence and significance. *Science* 191:404–06.

Diamond, J. M. 1978. Niche shifts and the rediscovery of interspecific competition. *Amer. Sci.* 66:322–31.

Cf. COMPETITIVE EXCLUSION; HABITAT SHIFT

NICHE-VARIATION HYPOTHESIS [of population diversity]
= niche width–variation hypothesis

Van Valen, L. 1965. Morphological variation and width of ecological niche. *Amer. Nat.* 99:377–90.

Soule, M., & B.R. Stewart. 1970. The "niche variation" hypothesis: a test and alternatives. *Amer. Nat.* 104:85–97.

Shugart, H. H., Jr., & B. G. Blaylock. 1973. The niche-variation hypothesis: an experimental study with *Drosophila* populations. *Amer. Nat.* 107:575–79.

Zangerl, A.R., S. T.A. Pickett, & F.A. Bazzaz. 1977. Some hypotheses on variation in plant populations and an experimental approach. *Biologist* 59:113–22.

Cf. LUDWIG'S HYPOTHESIS

NICHOLSON-BAILEY MODEL [of host-parasite interactions]

Nicholson, A.J., & V.A. Bailey. 1935. The balance of animal populations. I. *Proc. Zool. Soc. Lond.* 3:551–98.

Allen, J. C. 1975. Mathematical models of species interactions in time and space. *Amer. Nat.* 109:319–42.

Cf. HASSELL-VARLEY MODEL; LOTKA-VOLTERRA MODEL; TIME-SPACE MODEL

NICHOLSON'S THEORY [of insect population control]
= Nicholsonian control

Nicholson, A. J. 1933. The balance of animal populations. *J. Anim. Ecol.* 2:132–78.

______ . 1954. An outline of the dynamics of animal populations. *Aust. J. Zool.* 2:9–65.

Richards, O. W. 1961. The theoretical and practical study of natural insect populations. *Annu. Rev. Entomol.* 6:147–62.

Cf. DENSITY DEPENDENCE

NIELSEN'S POSTULATE [plankton density]

Nielsen, E. S. 1934. Untersuchungen über die Verbreitung, Biologie und Variation der Ceratien im südlichen stillen Ozean. *Dana-Rep.* 4:1–67.

Graham, H. W. 1941. An oceanographic consideration of the dinoflagellate genus *Ceratium*. *Ecol. Monogr.* 11:99–116.

NOAH'S ARK CASE [variability dispersal from an original population of two individuals]

Spiess, E. B. 1977. *Genes in Populations*. New York: Wiley, pp. 341*ff*.

Cf. BOTTLENECK EFFECT

NOMINALISTIC SPECIES CONCEPT *see* **SPECIES CONCEPTS**

NOMOGENESIS [evolution]

Berg, L.S. 1926. *Nomogenesis, or Evolution Determined by Law*. Transl. by J. N. Rostovtsow. London: Constable. (1969 reprint, Cambridge: M. I. T. Press.)

Dobzhansky, T. 1972. Darwinian evolution and the problem of extraterrestrial life. *Perspect. Biol. Med.* 15:157–75.

Cf. ORTHOGENESIS

NONADAPTIVE EVOLUTION

Van Valen, L. 1960. Nonadaptive aspects of evolution. *Amer. Nat.* 94:305–08.

NON-DARWINIAN EVOLUTION

King, J. L., & T.H. Jukes. 1969. Non Darwinian evolution. *Science* 164:788–98.

Richmond, R.C. 1970. Non-Darwinian evolution: a critique. *Nature* 225:1025–28.

Dobzhansky, T. 1972. Darwinian evolution and the problem of extraterrestrial life. *Perspect. Biol. Med.* 15:157–75.

Milkman, R. 1972. How much room is left for non-Darwinian evolution? *Brookh. Symp. Biol.* 23:217–29.

Cf. NEO-DARWINIAN EVOLUTION; NEUTRAL THEORY; RANDOM WALK; SELECTIONIST VIEW

NONDIMENSIONAL SPECIES CONCEPT *see* **SPECIES CONCEPTS**

NONEQUILIBRIUM HYPOTHESIS [island biogeography]

Brown, J. H. 1971. Mammals on mountaintops: nonequilibrium insular biogeography. *Amer. Nat.* 105:467–78.

Abbott, I., & P. R. Grant. 1976. Nonequilibrial bird faunas on islands. *Amer. Nat.* 110:507–28.

Cf. CONTEMPORANEOUS DISEQUILIBRIUM; EQUILIBRIUM THEORY [of island biogeography]; ISLAND BIOGEOGRAPHY, THEORY OF

 NONRECIPROCAL RECOMBINATION *see* **GENE CONVERSION**

NON-SEARCH IMAGE *see* **AVOIDANCE**

NON SYMBIOTIC THEORY [origin of eukaryotic cells]

Allsopp, A. 1969. Phylogenetic relationships of the procaryota and the origin of the eukaryotic cell. *New Phytol.* 68:591–612.

Raff, R. A., & H. R. Mahler. 1972. The non symbiotic origin of mitochondria. *Science* 177:575–82.

Cloud, P., M. Moorman, & D. Pierce. 1975. Sporulation and ultrastructure in a late Proterozoic cyanophyte: some implications for taxonomy and plant phylogeny. *Q. Rev. Biol.* 50:131–50.

Cf. SYMBIOTIC THEORY

NORMALIZING SELECTION [evolution]

= STABILIZING SELECTION = centripetal selection

Waddington, C. H. 1953. Epigenetics and evolution. *Symp. Soc. Exp. Biol.* 7:186–99.

Mather, K. 1955. Polymorphism as an outcome of disruptive selection. *Evolution* 9:52–61.

NUCLEAR THEORY OF RESPIRATION [physiology]

Verworn, M. 1892. Die physiologische Bedeutung des Zellkerns. *Pflügers Arch. Ges. Physiol.* 51:1–118.

Loeb, J. 1906. *The Dynamics of Living Matter*. New York: Columbia University Press.

Kohler, R E. 1973. The background to Otto Warburg's conception of the *Atmungsferment*. *J. Hist. Biol.* 6:171–92.

Cf. WARBURG'S MEMBRANE THEORY

NUCLEOPROTEIN THEORY [of the gene]

Olby, R. 1974. The origins of molecular genetics. *J. Hist. Biol.* 7:93–100.

NUCLEOSOME HYPOTHESIS [macronuclear organization in *Tetrahymena]*

Vorob'ev, V.I., S.N. Borehsenius, N.A. Belozerskaya, N. Merkulova, & I.S. Irlina. 1975. DNA replication in macronuclei of *Tetrahymena pyriformis* GL. *Exp. Cell Res.* 93:253–60.

Raikov, I.B. 1976. Evolution of macronuclear organization. *Annu. Rev. Genet.* 10:413–40.

Cf. SLAVE-REPEATS HYPOTHESIS; SUBNUCLEAR HYPOTHESIS

NUCLEOTYPE [cytogenetics]

Bennett, M.D. 1972. Nuclear DNA content and minimum generation time in herbaceous plants. *Proc. Roy. Soc. Lond. B* 181:109–35.

———. 1974. Nuclear characters in plants. *Brookh. Symp. Biol.* 23:344–66.

Stebbins, G. L. 1976. Chromosome DNA and plant evolution. *Evol. Biol.* 9:1–34.

NUNATAK HYPOTHESIS [biogeography]

Fernald, M.L. 1925. Persistence of plants in unglaciated areas of Boreal America. *Mem. Amer. Acad. Arts Sci.* 15: no. 3.

Wynne-Edwards, V. C. 1937. Isolated arctic-alpine floras in eastern North America: a discussion of their glacial and recent history. *Trans. Roy. Soc. Can.* 31:1–26.

Raup, H.M. 1941. Botanical problems in boreal America. I. *Bot. Rev.* 7:147–208.

Deevey, E.S., Jr. 1949. Biogeography of the Pleistocene. I. Europe and North America. *Bull. Geol. Soc. Amer.* 60:1315–416.

Cf. REFUGIUM THEORY

NUTRIENT RECOVERY HYPOTHESIS [of population cycles]

Pitelka, F.A. 1964. The nutrient-recovery hypothesis for arctic microtine cycles. I. *Grazing in Terrestrial and Marine Environments.* Edited by D. J. Crisp. *Brit. Ecol. Soc. Symp.* 4:55–56.

Schultz, A.M. 1964. The nutrient-recovery hypothesis for arctic microtine cycles. II. *Grazing in Terrestrial and Marine Environments.* Edited by D. J. Crisp. *Brit. Ecol. Soc. Symp.* 4:57–68.

———. 1969. A study of an ecosystem: the arctic tundra. *The Ecosystem Concept in Natural Resource Management.* Edited by G. Van Dyne. New York: Academic Press, pp. 77–93.

Pianka, E.R. 1978. *Evolutionary Ecology.* 2d ed. New York: Harper & Row, p. 124.

[O]

OCCAM'S RAZOR ["the simplest theory is preferred"]

Pearson, K. 1911. *The Grammar of Science.* London: Adam and Charles Black, p. 392.

Carre, M.H. 1946. *Realists and Nominalists.* Oxford: Oxford University Press, p. 107.

Carlson, E.A. 1966. *The Gene: A Critical History.* Philadelphia: Saunders, p. 251.

Hutchinson, G.E. 1978. *An Introduction to Population Ecology.* New Haven, Conn.: Yale University Press, pp. 2–3.

Cf. MORGAN'S CANON; PARSIMONY; SIMPLICITY

ODOR-TRAIL HYPOTHESIS [of bee recruitment]

Friesen, L.J. 1973. The search dynamics of recruited honey bees. *Biol. Bull.* 144:107–31.

Gould, J.L. 1976. The dance-language controversy. *Q. Rev. Biol.* 51:211–44.

OLD AGE THEORY [of retrogressive evolution]

Hyatt, A. 1866. On the parallelism between the different stages of the life in the individual and those in the entire group of the molluscous order Tetrabranchiata. *Mem. Boston Soc. Nat. Hist.* 1:193–209.

Gould, S.J. 1977. *Ontogeny and Phylogeny.* Cambridge: Belknap Press of Harvard University Press, pp. 93–94.

OLFACTION THEORY [of bee forager recruitment]

Frisch, K. von. 1919. Uber den Geruchesin der Beinen und seine blütenbiologische Bedeutung. *Zool. Jhrb. Abt. Allg. Zool. Physiol. Tiere* 37:1–238.

______. 1947. The dance of the honey bee. *Bull. Anim. Behav.* 5:1–32.

______. 1967. *The Dance Language and Orientation of Bees.* Cambridge: Harvard University Press, pp. 495*ff*.

Wenner, A.M. 1968. Honey bees. *Animal Communication.* Edited by T. A. Sebeok. Bloomington: Indiana University Press, pp. 217–43.

______, P.H. Wells, & D.L. Johnson. 1969. Honey bee recruitment to food sources: olfaction or language. *Science* 164:84–86.

Gould, J.L. 1976. The dance-language controversy. *Q. Rev. Biol.* 51:211–44.

OLFACTORY HYPOTHESIS [of salmon homing]

Hasler, A.D., & W.J. Wisby. 1951. Discrimination of stream odors by fishes and relation to parent stream behavior. *Amer. Nat.* 85:223–38.

Hasler, A.D., A.T. Scholz, & R. M. Horrall. 1978. Olfactory imprinting and homing in salmon. *Amer. Sci.* 66:347–55.

ONCOGENE HYPOTHESIS [of virus origin]

Huebner, R.I., & G.I. Todaro. 1969. Oncogenes of RNA tumor viruses as determinants of cancer. *Proc. Natl. Acad. Sci.* 64:1087–94.

Gross, L. 1974. Facts and theories on viruses causing cancer and leukemia. *Proc. Natl. Acad. Sci.* 71:2013–17.

Cf. PROTOVIRUS HYPOTHESIS

ONE BAND–ONE GENE HYPOTHESIS [genetics]

Lefevre, G., Jr. 1974. The one band–one gene hypothesis: evidence from a cytogenetic analysis of mutant and non-mutant rearrangement breakpoints in *Drosophila melanogaster*. *Cold Spr. Harb. Symp. Quant. Biol.* 38:591–99.

______ . 1974. The relationship between genes and polytene chromosome bands. *Annu. Rev. Genet.* 8:51–62.

ONE-CENTER EFFECT [in nerve growth]

Weiss, P. 1955. Nervous system (neurogenesis). *Analysis of Development.* Philadelphia: Saunders, pp. 346–401. (1971 reprint: New York: Hafner.)

Cf. TWO CENTER EFFECT

ONE GENE–ONE ENZYME HYPOTHESIS [genetics; biochemistry]

Beadle, G. W., & E. Tatum. 1941. Genetic control of biochemical reactions in Neurospora. *Proc. Natl. Acad. Sci.* 27:499–506.

Beadle, G. W. 1945. Genetics and metabolism in Neurospora. *Physiol. Rev.* 25:643–63.

______ . 1959. Genes and chemical reactions in Neurospora. *Science* 129:1715–19.

Horowitz, N.H., & V. Leupold. 1951. Some recent studies bearing on the one gene–one enzyme hypothesis. *Cold Spr. Harb. Symp. Quant. Biol.* 16:65–74.

Carlson, E. A. 1966. *The Gene: A Critical History.* Philadelphia: Saunders, pp. 166–73.

Hershey, A. D. 1970. Genes and hereditary characteristics. *Nature* 226:697–700.

ONE-LOCUS HYPOTHESIS [of chromosome "suicide"] *see* **OUTLAW GENES**

ONE PROTEIN HYPOTHESIS [genetics]

Bukhari, A. I. 1976. Bacteriophage MU as a transposition element. *Annu. Rev. Genet.* 10:389–412.

Cf. TWO PROTEINS HYPOTHESIS

ONE RESOURCE–ONE SPECIES PRINCIPLE [ecology]

MacArthur, R.H., & R. Levins. 1964. Competition, habitat selection, and character displacement in a patchy environment. *Proc. Natl. Acad. Sci.* 51:1207–10.

Rescigno, A., & I. W. Richardson. 1965. On the competitive exclusion principle. *Bull. Math. Biophys.* 27 (special issue): 85–89.

______ . 1967. The struggle for life. I. Two species. *Bull. Math. Biophys.* 29:377–88.

ONTOGENESIS [evolution]

Smit, P. 1962. Ontogenesis and phylogenesis: their interrelation and their interpretation. *Acta Biotheor.* 15:1–104.

Cf. BIOGENESIS; EPIGENESIS; RECAPITULATION; VON BAER'S LAWS

ONTOGENETICAL THEORY [of vertebrae origin]

Williams, E.E. 1959. Gadow's arcualis and the development of tetrapod vertebrae. *Q. Rev. Biol.* 34:1–32.

Dullemeijer, P. 1974. *Concepts and Approaches in Animal Morphology.* Assen, Netherlands: Van Gorcum, p. 16.

OOGENESIS-FLIGHT SYNDROME [ecology; biogeography]

Johnson, C.G. 1969. *Migration and Dispersal of Insects by Flight.* London: Methuen, chaps. 10, 25.

Baker, R.R. 1978. *The Evolutionary Ecology of Animal Migration.* New York: Holmes & Meier, chaps. 10, 17.

OPARIN-HALDANE HYPOTHESIS [of the origin of life]

Oparin, A.I. 1924. *Proiskhoyhdenie Zhizny.* Moscow: Izd. Moskovski Rabochü. (1936 ed. transl. as: *The Origin of Life,* by S. Morgulis. London: Macmillan, 1938.)

Haldane, J.B.S. 1929. The origin of life. *Rationalist Annual,* 1929.

Bernal, J.D. 1967. *The Origin of Life.* London: Weidenfeld & Nicolson, pp. 7, 24*ff.*

OPEN- and CLOSED-PROGRAM BEHAVIOR [acquired vs. instinctive behavior]

Mayr, E. 1974. Behaviour programs and evolutionary strategies. *Amer. Sci.* 62:650–59.

Boucot, A.J. 1978. Community evolution and rates of cladogenesis. *Evol. Biol.* 11:545–655.

OPERON THEORY [of gene action]

=Jacob-Monod operon theory

Jacob, F., & J. Monod. 1959. Gènes de structure and gènes de regolution dans la biosynthèse des proteines. *C. R. Acad. Sci.* 249:1282.

______. 1961. Genetic regulatory mechanisms in the synthesis of proteins. *J. Mol. Biol.* 3:318–56.

Jacob, F., D. Perrin, C. Sanchez, & J. Monod. 1960. L'opéron: groupe de gènes à expression coordonnée par un opérateur. *C.R. Acad. Sci.* 250:1727–29.

Ames, B.N., & R.C. Martin. 1964. Biochemical aspects of genetics: the operon. *Annu. Rev. Biochem.* 33:235–58.

Carlson, E.A. 1966. *The Gene: A Critical History.* Philadelphia: Saunders, pp. 221–30.

Buttin, G., F. Jacob, & J. Monod. 1967. The operon: a unit of coordinated gene action. *Heritage from Mendel.* Edited by R. A. Brink & E. D. Styles. Madison: University of Wisconsin Press, pp. 155–78.

Epstein, W., & J.R. Beckwith. 1968. Regulation of gene expression. *Annu. Rev. Biochem.* 37:411–36.

Reznikoff, W.S. 1972. The operon revisited. *Annu. Rev. Genet.* 6:133–56.

Schaffner, K.F. 1974. The peripherality of reductionism in the development of molecular biology. *J. Hist. Biol.* 7:111–39.

______. 1974. The unity of science and theory construction in molecular biology. *Bost. Stud. Philos. Sci.* 11:497–533.

OPTIMAL DIETS [ecology]

Pulliam, H.R. 1974. On the theory of optimal diets. *Amer. Nat.* 108:59–74.

OPTIMAL FORAGING THEORY [ecology]

MacArthur, R.H., & E.R. Pianka. 1966. On optimal use of a patchy environment. *Amer. Nat.* 100:603–09.

Schoener, T.W. 1971. Theory of feeding strategies. *Annu. Rev. Ecol. Syst.* 2:369–404.

Pearson, N.E. 1974. Optimal foraging theory. *Quant. Sci. Paper* no. 39, University of Washington, Seattle.

Charnov, E.L. 1976. Optimal foraging, the marginal value theorem. *Theor. Pop. Biol.* 9:129–36.

Parker, G.A., & R.A. Stuart. 1976. Animal behaviour as a strategy optimizer: evolution of resource assessment strategies and optimal emigration thresholds. *Amer. Nat.* 110:1055–76.

Pyke, G. H., H. R. Pulliam, & E. L. Charnov. 1977. Optimal foraging: a selective review of theory and tests. *Q. Rev. Biol.* 52:137–54.

OPTIMAL MANIFESTATION OF COLOR [in nature]

Peterich, L. 1972. Biological chromatology. The laws of colour and design in nature. *Acta Biotheor.* 21:24–46.

OPTIMALITY, PRINCIPLE OF

= economy in nature

Rosen, R. 1967. *Optimality Principles in Biology.* London: Butterworths.

Cf. PARSIMONY

OPTIMIZATION PRINCIPLE [genetic fitness]

Levins, R. 1968., Evolutionary consequences of flexibility. *Population Biology and Evolution.* Edited by R. C. Lewontin. Syracuse, N.Y.: Syracuse University Press, pp. 67–70.

OPTIMIZATION THEORY [behavior; evolution]

= optimality

Maynard Smith, J. 1978. Optimization theory in evolution. *Annu. Rev. Ecol. Syst.* 9:31–56.

Lewontin, R.C. 1979. Fitness, survival and optimality. *Analysis of Ecological Systems.* Edited by D. J. Horn, G. R. Stairs, & R. D. Mitchell. Columbus: Ohio State University Press, pp. 3–21.

Cf. ALTRUISM

OPTIMUM HETEROZYGOSITY HYPOTHESIS [genetics]

Mukai, T., S. Chigusa, & I. Yoshikawa. 1965. The genetic structure of natural populations of *Drosophila melanogaster.* III. Dominance effect of spontaneous mutant polygenes controlling viability in heterozygous genetic backgrounds. *Genetics* 52:493–501.

Mukai, T., & T. Yamazaki. 1968. The genetic structure of natural populations of *Drosophila melanogaster.* V. Coupling-repulsion effect of spontaneous mutant polygenes controlling viability. *Genetics* 59:513–35.

Mukai, T. 1969. The genetic structure of natural populations of *Drosophila melanogaster.* VI. Further studies on the optimum heterozygosity hypothesis. *Genetics* 61:479–95.

Cf. MAXIMUM HETEROZYGOSITY HYPOTHESIS

ORBITAL HYPOTHESIS [structural assembly in *Dictyostelium*]

Clark, R.L., & T.L. Steck. 1979. Morphogenesis in *Dictyostelium:* an orbital hypothesis. *Science* 204:1163–68.

ORDER [ecology; evolution; philosophy]

Pringle, J.W.S. 1951. On the parallel between learning and evolution. *Behaviour* 3:174–215.

Bertalanffy, L. von. 1952. *Problems of Life*. New York: Wiley, chaps. 4–5.

Bray, J.R. 1958. Notes toward an ecologic theory. *Evolution* 39:770–76.

ORGANIC CHANGE, THEORY OF [evolution]

Wallace, A.R. 1855. On the law which has regulated the introduction of new species. *Ann. Mag. Nat. Hist.* 16:184–96.

Brooks, J.L. 1972. Extinction and the origin of organic diversity. *Trans. Conn. Acad. Arts Sci.* 44:17–56.

ORGANIC SELECTION [evolution]

Weismann, A. 1894. *The Effect of External Influences upon Development*. Romanes Lecture. Oxford: Oxford University Press.

______ . 1909. The selection theory. *Darwin and Modern Science*. Edited by A. C. Seward. Cambridge: Cambridge University Press, pp. 18–65.

Baldwin, J.M. 1902. *Development and Evolution*. New York: Macmillan, chap. 8.

ORGANISM

Wright, S. 1953. Gene and organism. *Amer. Nat.* 87:5–18.

ORGANISMIC VIEW OF THE COMMUNITY *see* **SUPRAORGANISMIC CONCEPT**

ORGANIZATION TYPE *see* **TYPE** [morphological]

ORGANIZATIONAL DOMINANCE [ethology]

Walker, I., & R.M. Williams. 1976. The evolution of the cooperative group. *Acta Biotheor.* 25:1–43.

ORIANS-VERNER MODEL [ethology]

= polygyny threshold

Verner, J. 1964. Evolution of polygamy in the long-billed marsh wren. *Evolution* 18:252–61.

Orians, G.H. 1969. On the evolution of mating systems in birds and mammals. *Amer. Nat.* 103:589–603.

Wittenberger, J.F. 1976. The ecological factors selecting for polygyny in altricial birds. *Amer. Nat.* 110:779–99.

 ORIGINAL ANTIGENIC SIN, DOCTRINE OF [immunology]

Fayekas de St. Groth, S., & R.G. Webster. 1966. Disquisitions on original antigenic sin. I. Evidence in man. *J. Exp. Med.* 124:331–61.

Abramoff, P., & M.F. LaVie. 1970. *Biology of the Immune Response*. New York: McGraw-Hill, p. 237.

ORTHOGENESIS [evolution]

= Cartesian transformations = evolutionary trends = trends

Haacke, W. 1893. *Gestaltung und Vererbung*. Leipzig: Weigel.

Eimer, T. 1898. *On Orthogenesis*. Chicago: Open Court.

Jepsen, G.L. 1949. Selection, "orthogenesis," and the fossil record. *Proc. Amer. Philos. Soc.* 93:479–500.

Simpson, G.G. 1953. *The Major Features of Evolution*. New York: Columbia University Press, chap. 8.

Rensch, B. 1960. *Evolution above the Species Level*. New York: Columbia University Press, pp. 203*ff*.

Grant, V. 1977. *Organismic Evolution*. San Francisco: Freeman, pp. 259–61.

Cf. NOMOGENESIS

ORTHOGENETIC SPECIATION

Simpson, G. G. 1954. *The Meaning of Evolution*. New York: New American Library, pp. 30–51.

Vorzimmer, P. 1965. Darwin's ecology and its influence upon his theory. *Isis* 56:148–55.

ORTHOGENETIC THEORY [of bison evolution]

Schulz, B. C., & W. P. Frankforter. 1946. A geologic history of the bison in the Great Plains. *Bull. Univ. Nebr. State Mus.* 3:1–10.

Guthrie, R. D. 1970. Bison evolution and zoogeography in North America during the Pleistocene. *Q. Rev. Biol.* 45:1–15.

OSCILLATIONS [variation in size of populations]

Nicholson, A. J., & V. A. Bailey. 1935. The balance of animal populations. Part I. *Proc. Zool. Soc. Lond.* 3:551–98.

Allee, W. C., A. E. Emerson, O. Park, T. Park, & K. P. Schmidt. 1949. *Principles of Animal Ecology*. Philadelphia: Saunders, p. 837.

Carson, H. L. 1968. The population flush and its consequences. *Population Biology and Evolution*. Edited by R. C. Lewontin. Syracuse, N.Y.: Syracuse University Press, pp. 123–37.

Cf. FOUNDER-FLUSH SPECIATION THEORY

OUTGROWTH THEORY [nerve fiber development]

Bidder, F., & C. von Kupffer. 1857. *Untersuchungen über die Textur des Rückenmarks und die Entwickelung seiner Formelemente*. Leipzig: Breitkopf & Härtel.

Detwiler, S. R. 1933. Experimental studies upon the development of the amphibian nervous system. *Biol. Rev.* 8:269–310.

Harrison, R. G. 1969. *Organization and Development of the Embryo*. New Haven, Conn.: Yale University Press, p. 122.

Billings, S. M. 1971. Concepts of nerve fiber development, 1839–1930. *J. Hist. Biol.* 4:275–305.

Cf. MULTICELLULAR THEORY; NEURON THEORY; PROTOPLASMIC BRIDGE THEORY

OUTLAW GENES [genetics; evolution]

= outlaw concept

White, M. J. D. 1966. A case of spontaneous chromosome breakage at a specific locus occurring at meiosis. *Aust. J. Zool.* 14:1027–34.

Alexander, R. D., & G. Borgia. 1978. Group selection, altruism, and the levels of organization of life. *Annu. Rev. Ecol. Syst.* 9:449–74.

OVERDOMINANCE [genetics]

East, E. M. 1936. Heterosis. *Genetics* 21:375–97.

Hull, F. H. 1946. Overdominance and corn breeding where hybrid seed is not feasible. *J. Amer. Soc. Agron.* 38:1100–03.

———. 1952. Recurrent selection and overdominance. *Heterosis*. Edited by J. W. Gowen. Ames: Iowa State College Press, chap. 28.

Crow, J. F. 1952. Dominance and overdominance. *Heterosis*. Edited by J. W. Gowen. Ames: Iowa State College Press, chap. 18.

Shull, G. H. 1952. Beginnings of the heterosis concept. *Heterosis*. Edited by J. W. Gowen. Ames: Iowa State College Press, chap. 2.

OVERKILL HYPOTHESIS [biogeography]

Martin, P. S. 1967. Prehistoric overkill. *Pleistocene Extinctions: The Search for a Cause*. Edited by P. S. Martin & H. E. Wright. New Haven, Conn.: Yale University Press, pp. 75–120.

———. 1973. The discovery of America. *Science* 179:969–74.

OVERLAPPING GENES HYPOTHESIS [genetics]

Grasse, P. P. 1977. *Evolution of Living Organisms*. New York: Academic Press, p. 235.

Barrell, B. G., G. M. Air, & C. A. Hutchinson. 1976. Overlapping genes in bacteriophage $\phi X174$. *Nature* 264:34–41.

OVERLAPPING TRAITS HYPOTHESIS [genetics]

Ursell, J. H. 1977. A new interpretation of recombination frequencies in genetics. *Proceedings of the International Conference on Quantitative Genetics,* Aug. 16–21, 1976. Edited by E. Pollack, O. Kempthorne, & T. B. Bailey. Ames: Iowa State University Press, pp. 869–72.

OVERPRINTING HYPOTHESIS [of the gene]

Grasse, P. P. 1977. *Evolution of Living Organisms*. New York: Academic Press, p. 236.

OVERTON-MEYER CONCEPT [of axon membrane]

Overton, E. 1902. Beiträge zur allgemeinen Muskel- und Nervenphysiologie. *Pflügers Arch. Ges. Physiol.* 92:115–280.

Goldman, D. E. 1971. Excitability models. *Biophysics and Physiology of Excitable Membranes*. Edited by W. J. Adelman, Jr. New York: Van Nostrand Reinhold, pp. 337–58.

OWEN'S MODEL [of a genetical system]

Owen, A. R. G. 1952. A genetical system admitting of two stable equilibria. *Nature* 170:1127.

______. 1953. A genetical system admitting of two distinct stable equilibria under natural selection. *Heredity* 7:97–102.

Bodmer, W. F. 1965. Differential fertility in population genetics models. *Genetics* 51:411–24.

Mandel, S. P. H. 1971. Owen's model of a genetical system with differential viability between sexes. *Heredity* 26:49–63.

OXIDATION-REDUCTION HYPOTHESIS [of active transport]

Conway, E. J. 1957. Nature and significance of concentration relations of potassium and sodium ions in skeletal muscle. *Physiol. Rev.* 37:84–132.

PAEDOMORPHOSIS [morphogenesis]

Garstang, W. 1928. The morphology of the Tunicata, and its bearing on the phylogeny of the Chordata. *Q. J. Microsc. Sci.* 72:51–187.

Huxley, J. S. 1954. The evolutionary process. *Evolution as a Process*. Edited by J. Huxley, A. C. Hardy, & E. B. Ford. London: Allen & Unwin, pp. 1–23.

Carlquist, S. 1962. A theory of paedomorphosis in dicotyledonous woods. *Phytomorphology* 12:30–45.

Cf. NEOTENY

PAINT-POT THEORY OF HEREDITY *see* **BLENDING INHERITANCE**

PALINGENESIS [evolution]

Haeckel, E. 1866. *Generelle Morphologie der Organismen*. 2 vols. Berlin: Reimer.

George, T. N. 1933. Palingenesis and paleontology. *Biol. Rev.* 8:107–35.

de Beer, G. R. 1958. *Embryos and Ancestors*. 3d ed. Oxford: Clarendon Press, chap. 1.

Cf. BIOGENETIC LAW

PANBIOGEOGRAPHY [biogeography]

Croizat, L. 1958. *Panbiogeography*. 2 vols. Caracas: By the author.

Nelson, G. 1973. Comments on Leon Croizat's biogeography. *Syst. Zool.* 22:312–20.

Croizat-Chaley, L. 1976. Biogeografia Analitica y Sintetica ("Panbiogeografia") de las Americas. *Biblioteca de la Academia de Ciencias Fisicas, Matematicas y Naturales, Caracas* 15:1–454; 16:455–890.

Cf. GENERALIZED TRACKS

PANGENESIS [heredity]

Darwin, C. 1868. *The Variation of Animals and Plants Under Domestication*. 2 vols. London: Murray, vol. 2, chap. 27.

Brooks, W. K. 1877. A provisional hypothesis of pangenesis. *Amer. Nat.* 11:144–47.

deVries, H. 1910. *Intracellular Pangenesis*. Transl. by C. S. Gager. Chicago: Open Court.

Zirkle, C. 1946. Early history of the idea of inheritance of acquired characters and of pangenesis. *Trans. Amer. Philos. Soc.* 35:91–151.

Olby, R. C. 1963. Charles Darwin's manuscript of pangenesis. *Brit. J. Hist. Sci.* 1:258–59. Reprinted in *Origins of Mendelism*, by R. C. Olby. New York: Schocken, 1966, pp. 173–75.

Geison, G. L. 1969. Darwin and heredity: the evolution of his hypothesis of pangenesis. *J. Hist. Med.* 24:375–411.

Ghiselins, M. T. 1975. The rationale of pangenesis. *Genetics* 79 (Suppl.): 47–57.

Pollock, M. R. 1976. From pangens to polynucleotides: the evolution of ideas on the mechanism of biological replication. *Perspect. Biol. Med.* 19:455–72.

PANMIXIA *see* **ACCUMULATION OF RANDOM MUTATIONS, THEORY OF**

PANSPERMIA HYPOTHESIS [origin of life]

Oparin, A. I. 1957. *The Origin of Life on the Earth*. New York: Academic Press, chap. 2.

______. 1968. *Genesis and Evolutionary Development of Life*. New York: Academic Press, p. 28.

PARADOX OF ENRICHMENT [ecology]

Rosenzweig, M. L. 1972. Stability of enriched aquatic ecosystems. *Science* 175:564–65.

Roughgarden, J. 1979. *Theory of Population Genetics and Evolutionary Ecology: An Introduction*. New York: Macmillan, p. 447.

PARADOX OF THE PLANKTON [coexistence of closely competing species]

Hutchinson, G. E. 1961. The paradox of the plankton. *Amer. Nat.* 95:137–45.

Richerson, P., R. Armstrong, & C. R. Goldman. 1970. Contemporaneous disequilibrium, a new hypothesis to explain the "paradox of the plankton." *Proc. Natl. Acad. Sci.* 67:1710–14.

Phillips, O. M. 1973. The equilibrium and stability of simple marine biological systems. I. Primary nutrient consumers. *Amer. Nat.* 107:73–93.

Stewart, F. M., & B. R. Levin. 1973. Partitioning of resources and the outcome of interspecific competition: a model and some general considerations. *Amer. Nat.* 107:171–98.

Petersen, R. 1975. The paradox of the plankton: an equilibrium hypothesis. *Amer. Nat.* 109:35–49.

PARALLEL EVOLUTION
= PARALLELISM

Naudin, C. 1856. Nouvelles recherches sur les caractères spécifiques et les variétés des plantes du genre *Cucurbita*. *Ann. Sci. Nat.*, ser. 4, 6:5–73.

Arber, A. 1925. *Monocotyledons: A Morphological Study*. Cambridge: Cambridge University Press, chap. 10.

Simpson, G. G. 1961. *Principles of Animal Taxonomy*. New York: Columbia University Press, pp. 103*ff*.

Davis, P. H. & V. H. Heywood. 1965. *Principles of Angiosperm Taxonomy*. Edinburgh: Oliver & Boyd, p. 42.

Went, F. W. 1971. Parallel evolution. *Taxon* 20:197–226.

Peet, R. K. 1978. Ecosystem convergence. *Amer. Nat.* 112:441–44.

Cf. CONVERGENT EVOLUTION

PARALLELISM [in morphology or evolution]
= PARALLEL EVOLUTION

Zawarzin, A. 1925. Der Parallelismus der Strukturen als ein Grundprinzip der Morphologie. *Z. Wiss. Zool.* 124:118–212.

Arber, A. 1950. *The Natural Philosophy of Plant Form*. Cambridge: Cambridge University Press, p. 159.

PARAMUTATION [genetics]

Hagemann, R. 1969. Somatic conversion (paramutation) at the *sulfurea* locus of *Lycopersicom esculentum* Mill. III. Studies with trisomics. *Can. J. Genet. Cytol.* 11:346–58.

Brink, R. A., E. D. Styles, & J. D. Axtell. 1968. Paramutation: directed genetic change. *Science* 159:161–70.

Brink, R. A. 1973. Paramutation. *Annu. Rev. Genet.* 7:129–52.

PARANOTAL THEORY [of insect wing origin]
= flying squirrel theory

Crampton, G. C. 1916. The phylogenetic origin and the nature of the wings of insects according to the paranotal theory. *J. N.Y. Entomol. Soc.* 24:1–38.

Alexander, R. D., & W. L. Brown. 1963. Mating behavior and the origin of insect wings. *Occas. Pap. Mus. Zool. Univ. Mich.* no. 628:1–19.

Wigglesworth, V. B. 1976. The evolution of insect flight. *Insect Flight*. Edited by R. C. Rainey. Oxford: Blackwell. *Symp. Roy. Entomol. Soc. Lond.* 7:255–69.

Daly, H. V., J. T. Doyen, & P. R. Ehrlich. 1978. *Introduction to Insect Biology and Diversity*. New York: McGraw-Hill, p. 266.

Kukalova-Peck, J. 1978. Origin and evolution of insect wings and their relation to metamorphosis, as documented by the fossil record. *J. Morph.* 156:53–126.

Cf. GILL THEORY

PARAPATRIC SPECIATION

= STASIPATRIC SPECIATION

Murray, J. 1972. *Genetic Diversity and Natural Selection.* New York: Hafner, pp. 87*ff*.

Bush, G. L. 1975. Modes of animal speciation. *Annu. Rev. Ecol. Syst.* 6:339–61.

PARAPHYLESIS *see* **PARAPHYLY**

PARAPHYLY [evolution]

Hennig, W. 1966. *Phylogenetic Systematics.* Urbana: University of Illinois Press, chap. 2.

Ashlock, P. D. 1971. Monophyly and associated terms. *Syst. Zool.* 20:63–69.

Nelson, G. J. 1971. Paraphyly and polyphyly: redefinitions. *Syst. Zool.* 20:471–72.

PARENT-OFFSPRING CONFLICT THEORY [behavior]

Trivers, R. L. 1974. Parent-offspring conflict. *Amer. Zool.* 14:249–64.

Lewontin, R. C. 1979. Fitness, survival and optimality. *Analysis of Ecological Systems.* Edited by D. H. Horn, G. R. Stairs, & R. D. Mitchell. Columbus: Ohio State University Press, pp. 3–21.

PARENTAL INVESTMENT, THEORY OF [behavior]

Trivers, R. L. 1972. Parental investment and sexual selection. *Sexual Selection and the Descent of Man.* Edited by B. G. Campbell. Chicago: Aldine, pp. 136–79.

Wilson, E. O. 1975. *Sociobiology.* Cambridge: Belknap Press of Harvard University Press, pp. 324–27.

Smith, J. M. 1977. Parental investment: a prospective analysis. *Anim. Behav.* 25:1–9.

PARSIMONY [simplicity in nature]

Hamilton, W. 1852. *Discussions on Philosophy and Literature.* London: Longman, Brown, Green & Longmans, p. 590.

Pearson, K. 1911. *The Grammar of Science.* London: Adam and Charles Black, p. 392.

Kohlberger, W. 1978. Footnote in review of: *Problems in Vertebrate Evolution.* Edited by S. Mahala Andrews, R. S. Miles, & A. D. Walker. New York & London: Academic Press. *Syst. Zool.* 27:373–78.

Cf. MINIMUM EVOLUTION; OCCAM'S RAZOR; SIMPLICITY

PARTIAL CHIASMATYPE HYPOTHESIS *see* **CHIASMATYPE THEORY**

PARTIAL EQUIVALENCE, PRINCIPLE OF [ecology]

Allee, W. C., A. C. Emerson, O. Park, T. Park, & K. P. Schmidt. 1949. *Principles of Animal Ecology*. Philadelphia: Saunders, pp. 223–24.

Cf. MINIMUM, LAW OF

PARTIAL-SHOOT THEORY [leaf morphology]

Candolle, C. P. de. 1868. Théorie de la feuille. *Arch. Sci. Phys. Nat.* 32:32–64.

Arber, A. 1941. The interpretation of leaf and root in the angiosperms. *Biol. Rev.* 16:81–105.

______ . 1950. *The Natural Philosophy of Plant Form*. Cambridge: Cambridge University Press, pp. 70–92*ff*.

PARTICULATE CONCEPT OF HEREDITY

Fisher, R. A. 1918. The correlation between relatives on the supposition of Mendelian inheritance. *Trans. Roy. Soc. Edinb.* 52:399–433.

______ . 1930. *The Genetical Theory of Natural Selection*. Oxford: Clarendon Press, Chap. 1.

Dobzhansky, T. 1951. *Genetics and the Origin of Species*. 3d ed. New York: Columbia University Press, pp. 51–52.

Cf. CHROMOSOME THEORY; BLENDING CONCEPT

PASTEUR EFFECT or REACTION [regulation of respiration]

Warburg, O. 1926. Über die Wirkung von Blausäureäthylester (Äthylcarbylamin) auf die Pasteurische Reaktion. *Biochem. Z.* 172:432–41.

Krebs, H. A. 1972. The Pasteur effect and the relations between respiration and fermentation. *Essays in Biochemistry*. Edited by P. N. Campbell & F. Dickens. New York: Academic Press, 8:1–34.

Racker, E. 1974. History of the Pasteur effect and its pathobiology. *Molec. Cell. Biochem.* 5:17–23.

Cf. CRABTREE EFFECT

PATCHY ENVIRONMENT [spatial properties of an environment]

= environmental patchiness

MacArthur, R., & R. Levins. 1964. Competition, habitat

selection, and character displacement in a patchy environment. *Proc. Natl. Acad. Sci.* 51:1207–10.

MacArthur, R., & E. Pianka. 1966. On optimal use of a patchy environment. *Amer. Nat.* 100:603–10.

Gillespie, J. 1974. Polymorphism in patchy environments. *Amer. Nat.* 108:145–51.

Levin, S. A., & R. T. Paine. 1974. Disturbance, patch formation, and community structure. *Proc. Natl. Acad. Sci.* 71:2744–47.

PATHOGEOGRAPHY *see* **GEOPHYTOPATHOLOGY**

PATTERN [in ecology]

Hutchinson, G. E. 1953. The concept of pattern in ecology. *Acad. Nat. Sci. Phila. Proc.* 105:1–12.

PATTERN EFFECT [gene action]

Goldschmidt, R. B. 1955. *Theoretical Genetics.* Berkeley: University of California Press, p. 398.

Grant, V. 1964. *The Architecture of the Germplasm.* New York: Wiley, chaps. 6 & 8.

Cf. PLEIOTROPHY

PELTATE CARPEL THEORY [origin of the carpel]

Čelakovský, L. 1876. Vergleichende Darstellung der Plazenten im Fruchtknoten der Phanerogamen. *Abh. Königl. Böhm. Ges. Wiss.* 6 Folge 8.

Troll, W. 1932. Morphologie der schildformigen Blätter. *Planta* 17:153–314.

Franck, D. H. 1976. The morphological interpretation of epiascidate leaves. *Bot. Rev.* 42:345–88.

PELTATION THEORY [of the tubular leaf on *Nepenthes]*

Goebel, K. 1900. *Organography of Plants.* Part I. Oxford: Clarendon Press.

Troll, W. 1939. *Vergleichende Morphologie der höheren Pflanzen.* Band I. *Vegetationsorgane,* teil 2. Berlin: Gebrüder Borntraeger.

Franck, D. H. 1976. The morphological interpretation of epiascidate leaves. *Bot. Rev.* 42:345–88.

PENETRATION HYPOTHESIS [microbiology]

Delbrück, M. 1945. Interference between bacterial viruses. III. The mutual exclusion effect and the depressor effect. *J. Bacteriol.* 50:151–70.

PENINSULAR EFFECT [animal distribution]

Simpson, G. G. 1964. Species density of North American recent mammals. *Syst. Zool.* 13:57–73.

Taylor, R. J., & P. J. Regal. 1978. The peninsular effect on species diversity and the biogeography of Baja California. *Amer. Nat.* 112:583–93.

PERFECT DENSITY DEPENDENCE [insect population control]

Milne, A. 1957. The natural control of insect populations. *Can. Entomol.* 89:193–213.

______ . 1957. Theories of natural control of insect populations. *Cold Spr. Harb. Symp. Quant. Biol.* 22:253–71.

______ . 1962. On a theory of natural control of insect population. *J. Theor. Biol.* 3:19–50.

Cf. NATURAL CONTROL, THEORIES OF; NICHOLSON'S THEORY

PERICAULOME THEORY [morphology]

Potonie, H. 1902. Die Pericaulom-Theorie. *Ber. Deut. Bot. Ges. 20:502–20.*

Arber, A. 1930. Root and shoot in the angiosperms: a study of morphological categories. *New Phytol.* 29:297–315.

Cf. LEAF-SKIN THEORY

PHAGOCYTOSIS THEORY [of immunity]

Metchnikoff, E. 1884. Ueber eine Sprosspilzkrankheit der Daphnien. Beitrag zur Lehre über den Kampf der Phagocyten gegen Krankheitserreger. *Arch. Path. Anat.* 96:177–95.

Bulloch, W. 1938. *The History of Bacteriology*. Oxford: Oxford University Press, chap. 11.

PHASE THEORY [locust population density]

= phase transformation theory

Uvarov, B. P. 1921. A revision of the genus *Locusta* L. (= *Pacytylus* Fiele), with a new theory as to the periodicity and migrations of locusts. *Bull. Entomol. Res.* 12:135–63.

______ . 1928. *Locusts and Grasshoppers. A Handbook for their Study and Control.* London: Imperial Bureau of Entomology.

Key, K. H. L. 1950. A critique of the phase theory of locusts. *Q. Rev. Biol.* 25:363–407.

Kennedy, J. S. 1956. Phase transformation in locust biology. *Biol. Rev.* 31:349–70.

Nolte, D. J. 1974. The gregarization of locusts. *Biol. Rev.* 49:1–14.

PHENETIC SYSTEMATICS [taxonomy]

Sokal, R. R., & P. H. A. Sneath. 1963. *Principles of Numerical Taxonomy*. San Francisco: Freeman, pp. 20*ff*, 90*ff*.

Hull, D. L. 1970. Contemporary systematic philosophies. *Annu. Rev. Ecol. Syst.* 1:19-54.

Sneath, P. H. A. 1971. Numerical taxonomy: criticisms and critiques. *Biol. J. Linn. Soc.* 3:147–57.

______ , & R. R. Sokal. 1973. *Numerical Taxonomy*. San Francisco: Freeman, p. 9*ff*.

Cf. EVOLUTIONARY SYSTEMATICS; PHYLOGENETIC SYSTEMATICS

PHENOLOGY

Lieth, H. 1970. Phenology in productivity studies. *Analysis of Temperate Forest Ecosystems*. Edited by D. E. Reichle. New York: Springer-Verlag, pp. 29–46.

LE PHENOMÈNE DE DANYSZ *see* **EHRLICH'SCHE PHÄNOMEN**

PHENON [taxonomy]

Sneath, P. H. A., & R. R. Sokal. 1962. Numerical taxonomy. *Nature* 193:855–60.

______ . 1973. *Numerical Taxonomy*. San Francisco: Freeman, p. 294.

PHENOTYPE [genetics]

Johannsen, W. 1909. *Elemente der exakten Erblichkeitslehre*. Jena: Fischer.

Turesson, G. 1922. The species and the variety as ecological units. *Hereditas* 3:100–13.

Churchill, F. B. 1974. William Johannsen and the genotype concept. *J. Hist. Biol.* 7:5–30.

Wanscher, J. H. 1975. The history of William Johannsen's genetical terms and concepts from the period 1903 to 1926. *Centaurus* 19:125–47.

PHENOTYPIC ADAPTATION [clutch-sizes of birds]

Lack, D. 1954. *The Natural Regulation of Animal Numbers*. Oxford: Clarendon Press, chap. 4.

PHENOTYPIC FLEXIBILITY *see* **DEVELOPMENTAL HOMEOSTASIS**

PHENOTYPIC PLASTICITY [morphogenesis]

= plasticity of the phenotype

Ganong, W. F. 1901. The cardinal principles of morphology. *Bot. Gaz.* 31:426–34.

Bradshaw, A. D. 1965. Evolutionary significance of phenotypic plasticity in plants. *Adv. Genet.* 13:115–56.

______ . 1974. Environment and phenotypic plasticity. *Brookh. Symp. Biol.* 25:75–94.

Marshall, D. R., & S. K. Jain. 1968. Phenotypic plasticity of *Avena fatua* and *A. barbata*. *Amer. Nat.* 102:457–67.

Cf. DEVELOPMENTAL HOMEOSTASIS

PHENOTYPIC SPECIES CONCEPT *see* **SPECIES CONCEPTS**

PHOTOPERIODISM [plant physiology]

Garner, W. W., & H. A. Allard. 1920. Effect of the relative length of day and night and other factors of the environment on growth and reproduction in plants. *J. Agric. Res.* 18:553–606.

______ . 1923. Further studies in photoperiodism, the response of the plant to relative length of day and night. *J. Agric. Res.* 23:871–920.

Murneek, A. E. 1948. History of research in photoperiodism. *Vernalization and Photoperiodism: A Symposium.* Edited by A. E. Murneek & R. O. Whyte. Waltham, Mass.: Chronica Botanica, pp. 39–61.

Evans, L. T. 1969. A short history of the physiology of flowering. *The Induction of Flowering: Some Case Histories.* Edited by L. T. Evans. Ithaca, N.Y.: Cornell University Press, pp. 1–13.

Chailakhyan, M. K. 1975. Forty years of research on the hormonal basis of plant development—some personal reflections. (Introd. by A. D. Krikorian.) *Bot. Rev.* 41:1–29.

PHYLETIC GRADUALISM [evolution]

= phyletic evolution

Simpson, G. G. 1944. *Tempo and Mode in Evolution.* New York: Columbia University Press, chap. 7.

______ . 1953. *The Major Features of Evolution.* New York: Columbia University Press, chap. 12.

______ . 1970. Uniformitarianism. An inquiry into principle, theory and method in geohistory and biohistory. *Essays in Evolution and Genetics in Honor of Theodosius Dobzhansky.* Edited by M. K. Hecht & W. C. Steere. New York: Appleton-Century-Crofts, pp. 43–96.

Pijl, L. van der. 1960. Ecological aspects of flower evolution. I. Phyletic evolution. *Evolution* 14:403–16.

Eldredge, N., & S. J. Gould. 1972. Punctuated equilibria: an alternative to phyletic gradualism. *Models in Paleobiology*. Edited by T. J. M. Schopf. San Francisco: Freeman, Cooper, pp. 82–115.

Cf. ANAGENESIS; GEOGRAPHICAL SPECIATION; QUANTUM EVOLUTION; SALTATION; SPECIES SELECTION

PHYLLEMBRYOGENESIS, THEORY OF [relations between ontogeny and phylogeny]

= phylembryogenesis

Severtsov, A. N. 1931. *Morphologische Gesetzmässigkeiten der Evolution*. Jena: Fischer.

Takhtajan, A. L. 1945. An essay of application of the theory of phyllembryogenesis to the interpretation of monocotyledonous embryo. *Proc. Acad. Sci. Armenian SSR* 3:53.

Hennig, W. 1966. *Phlogenetic Systematics*. Urbana: University of Illinois Press, p. 96.

PHYLLODE THEORY [leaf origin]

Candolle, A. P. de. 1827. *Organographie*. Tome I. Paris: Germer Bailliere.

Arber, A. 1918. The phyllode theory of the monocotyledonous leaf, with special reference to anatomical evidence. *Ann. Bot.* 32:465–501.

______. 1921. The leaf structure of the *Iridaceae*, considered in relation to the phyllode theory. *Ann. Bot.* 35:301–36.

Kaplan, D. R. 1973. The monocotyledons: their evolution and comparative biology. VII. The problem of leaf morphology and evolution in the monocotyledons. *Q. Rev. Biol.* 48:437–57.

Franck, D. H. 1976. The morphological interpretation of epiascidate leaves. *Bot. Rev.* 42:345–88.

PHYLLODIAL THEORY [origin of the tubular leaf of *Nepenthes*]

Candolle, A. P. de. 1827. *Organographie*. Tome I. Paris: Germer Bailliere.

Troll, W. 1939. *Vergleichende Morphologie der Höheren Pflanzen*. Band I. *Vegetationsorgane*, teil 2. Berlin: Gebrüder Borntraeger.

Franck, D. H. 1976. The morphological interpretation of epiascidate leaves. *Bot. Rev.* 42:345–88.

PHYLLOTAXIS or PHYLLOTAXY [leaf arrangement]
= Goethe's spiral tendency = spiral theory

Church, A. H. 1904. *On the Relation of Phyllotaxis to Mechanical Laws*. London: Williams & Norgate.

Thompson, D'A. W. 1942. *On Growth and Form*. 2d ed. Cambridge: Cambridge University Press, chap. 14.

Snow, M., & R. Snow. 1934. The interpretation of phyllotaxis. *Biol. Rev.* 9:132–37.

Snow, R. 1955. Problems of phyllotaxis and leaf determination. *Endeavour* 14:190–99.

Leppik, E. E. 1961. Phyllotaxis, anthotaxis and semataxis. *Acta Biotheor.* 14:1–28.

Montgomery, W. M. 1970. The origins of the spiral theory of phyllotaxis. *J. Hist. Biol.* 3:299–323.

PHYLOGENESIS [evolution]

Smit, P. 1962. Ontogenesis and phylogenesis: their interrelation and their interpretation. *Acta Biotheor.* 15:1–104.

Hennig, W. 1966. *Phylogenetic Systematics*. Urbana: University of Illinois Press.

PHYLOGENETIC BRAIN PROGRESS, RULE OF [evolution]

Edinger, T. 1948. Evolution of the horse brain. *Mem. Geol. Soc. Amer.* no. 25.

Rensch, B. 1967. The evolution of brain achievements. *Evol. Biol.* 1:26–68.

PHYLOGENETIC BRANCHING *see* **CLADOGENESIS**

PHYLOGENETIC INTERMEDIATE RULE [evolution; systematics]

Hennig, W. 1966. The Diptera fauna of New Zealand as a problem in systematics and zoogeography. Transl. by P. Wygodzinsky. *Pac. Insects Monogr.*, no. 9:1–81.

Ashlock, P. D. 1974. The uses of cladistics. *Annu. Rev. Ecol. Syst.* 5:81–99.

PHYLOGENETIC SCALE *see* **CHAIN OF BEING; SCALA NATURAE**

PHYLOGENETIC SPECIES CONCEPT *see* **SPECIES CONCEPTS**

PHYLOGENETIC SYSTEMATICS
= cladistics

Brundin, L. 1966. Transantarctic relationships and their significance, as evidenced by chironomid midges. *K. Sven. Vetenskapsakad. Handl.*, ser. 4, vol. 11, no. 1.

Hennig, W. 1966. *Phylogenetic Systematics.* Urbana: University of Illinois Press.

Darlington, P. J., Jr. 1970. A practical criticism of Hennig-Brundin "phylogenetic systematics" and Antarctic biogeography. *Syst. Zool.* 19:1–18.

Hull, D. L. 1970. Contemporary systematic philosophies. *Annu. Rev. Ecol. Syst.* 1:19–54.

Nelson, G. T. 1971. "Cladism" as a philosophy of classification. *Syst. Zool.* 20:373–76.

______. 1972. Comments on Hennig's "Phylogenetic Systematics" and its influence on ichthyology. *Syst. Zool.* 21:364–74.

Brundin, L. 1972. Evolution, causal biology, and classification. *Zool. Scr.* 1:107–20.

Kavanaugh, D. H. 1972. Hennig's principles and methods of phylogenetic systematics. *Biologist* 54:115–27.

Abdullah, M. 1973. The improvement of an existing modern classification in biology. *Zool. Beitr.* 19:13–41.

Mayr, E. 1974. Cladistic analysis or cladistic classification? *Z. Zool. Syst. Evolutionsforsch.* 12:94–128.

Bremer, K., & H. Wanntorp. 1978. Phylogenetic systematics in botany. *Taxon* 27:317–29.

Cf. EVOLUTIONARY SYSTEMATICS; PHENETIC SYSTEMATICS

PHYLOGENETIC TREE [taxonomy]

Arber, A. 1950. *The Natural Philosophy of Plant Form.* Cambridge: Cambridge University Press, pp. 65–66.

Hennig, W. 1966. *Phylogenetic Systematics.* Urbana: University of Illinois Press, chap. 2.

Cf. WAGNER DIVERGENCE INDEX

PHYLOGENY [evolution]

Crow, W. B. 1926. Phylogeny and the natural system. *J. Genet.* 17:85–155.

Hennig, W. 1966. *Phylogenetic Systematics.* Urbana: University of Illinois Press.

Leppik, E. E. 1974. Phylogeny, hologeny and coenogeny, basic concepts of environmental biology. *Acta Biotheor.* 23:170–93.

PHYSIOLOGICAL CLOCK *see* **BIOLOGICAL CLOCK**

PHYSIOLOGICAL HOMEOSTASIS [physiology]
= wisdom of the body

Cannon, W. B. 1929. Organization for physiological homeostatics. *Physiol. Rev.* 9:399–431.

Grant, V. 1963. *The Origin of Adaptations.* New York: Columbia University Press, pp. 122–23.

Dubos, R. 1965. *Man Adapting.* New Haven, Conn.: Yale University Press, pp. 256, 290*ff.*

Cf. MILIEU INTÉRIEUR

PHYSIOLOGICAL LIMITS, THEORY OF [plant geography]

Livingston, B. E., & E. Shreve. 1921. The distribution of vegetation in the United States as related to climatic conditions. *Carnegie Inst. Wash. Publ.* 284:1–585.

Mason, H. L. 1936. The principles of geographic distribution as applied to floral analysis. *Madroño* 3:181–90.

PHYSIOLOGICAL SELECTION [evolution]

Romanes, G. J. 1886. Physiological selection: an additional suggestion on the origin of species. *J. Linn. Soc. Lond. Zool.* 19:337–411.

Lesch, J. E. 1975. The role of isolation in evolution: George J. Romanes and John T. Gulick. *Isis* 66:483–503.

Cf. GEOGRAPHIC ISOLATION; REPRODUCTIVE ISOLATION

PHYTOALEXIN PRINCIPLE or THEORY [plant pathology]

Müller, K. O., & H. Börger. 1940. Experimentelle untersuchungen über die Phytophthora—Resistenz der Kartoffel. *Arb. Biol. Reichsanst. Landw. Forstw. Berlin* 23:189–231.

Müller, K. O. 1961. The phytoalexin concept and its methodological significance. *Recent Advances in Botany.* 2 vols. Toronto: University of Toronto Press, 1:396–400.

Cruickshank, I. A. M. 1963. Phytoalexins. *Ann. Rev. Phytopath.* 1:351–74.

Ingham, J. L. 1972. Phytoalexins and other natural products as factors in plant disease resistance. *Bot. Rev.* 38:343–424.

PHYTOCHROME [plant physiology]

Hendricks, S. B., & H. A. Borthwick. 1967. The function of phytochrome in the regulation of plant growth. *Proc. Natl. Acad. Sci.* 58:2125–30.

Mitrakos, K., & W. Shropshire, Jr. 1972. *Phytochrome.* New York: Academic Press.

PHYTOCHROME ONLY HYPOTHESIS [plant response to high photoirradiation]

Hendricks, S. B., & H. A. Borthwick. 1959. Photocontrol of plant development by the simultaneous excitation of two interconvertible pigments. *Proc. Natl. Acad. Sci.* 45:344–49.

Hartmann, K. M. 1966. A general hypothesis to interpret "high energy phenomena" of photomorphogenesis on the basis of phytochrome. *Photochem. Photobiol.* 5:349–66.

Mancinelli, A. L., & I. Rabino. 1978. The "high irradiance responses" of plant photomorphogenesis. *Bot. Rev.* 44:129–80.

PHYTOCHROME PLUS HYPOTHESIS [plant response to high photoirradiation]

Mancinelli, A. L., & I. Rabino. 1978. The "high irradiance responses" of plant photomorphogenesis. *Bot. Rev.* 44:129–80.

PHYTONIC THEORY [of shoot organization in vascular plants]
= phyton hypothesis

Spencer, H. 1864–67. *The Principles of Biology*. 2 vol. London: Williams & Norgate, vol. 1, p. 479; vol. 2, p. 8.

Wetmore, R. H. 1943. Leaf-stem relationships in the vascular plants. *Torreya* 43:16–28.

Majumdar, G. P. 1947. Growth unit or the phyton in dicotyledons with special reference to *Heracleum*. *Bull. Bot. Soc. Bengal*, April, 1947, pp. 61–66.

Arber, A. 1950. *The Natural Philosophy of Plant Form*. Cambridge: Cambridge University Press, pp. 166*ff*.

Wardlaw, C. W. 1952. *Phylogeny and Morphogenesis*. London: Macmillan, p. 427.

Cf. AXIAL THEORY; BIOLOGICAL MAINTENANCE

PICNIC HYPOTHESIS [origin of life on earth]
= garbage hypothesis

Gold, T. 1960. Cosmic garbage. *Air Force and Space Digest* May, 1960, p. 65.

PIPE MODEL THEORY [of tree form]

Shinozaki, K., K. Yoda, K. Hozumi, & T. Kira. 1964. A quantitative analysis of plant form—the pipe model theory. I. Basic analysis. II. Further evidence of the theory and its application in forest ecology. *Jap. J. Ecol.* 14:97–105, 133–39.

PLANT DEFENSE GUILDS *see* **GUILD DEFENSE**

PLASMAGENE [cytoplasmic hereditary factor]
Darlington, C. D. 1944. Heredity, development and infection. *Nature* 154:164–69.
Carlson, E. A. 1966. *The Gene: A Critical History*. Philadelphia: Saunders, pp. 174–83.

PLASMODESM THEORY [of nerve development]
= PROTOPLASMIC BRIDGE THEORY
Hensen, V. 1864. Über die Entwickelung des Gewebes und der Nerven im Schwanze der Froschlarve. *Virchows Arch. Path. Anat.* 31:51–73.
Detwiler, S. 1933. Experimental studies upon the development of the amphibian nervous system. *Biol. Rev.* 8:269–310.
Harrison, R. G. 1969. *Organization and Development of the Embryo*. New Haven, Conn.: Yale University Press, p. 122.

PLASTICITY OF THE PHENOTYPE *see* **PHENOTYPIC PLASTICITY**

PLATE TECTONICS [biogeography]
Dewey, J. F., & B. Horsfield. 1970. Plate tectonics, orogeny and continental growth. *Nature* 225:521–25.
McKenzie, D. P. 1970. Plate tectonics and continental drift. *Endeavor* 29:39–44.
Schuster, R. M. 1976. Plate tectonics and its bearing on the geographical origin and dispersal of angiosperms. *Origin and Early Evolution of Angiosperms*. Edited by C. B. Beck. New York: Columbia University Press, pp. 48–138.
Cf. CONTINENTAL DRIFT

PLEIOTROPHY [genetics]
= multiple gene effects
Plate, L. 1910. *Vererbungslehre und Deszendenztheorie*. Festschrift für R. Hertwig. II. Jena: Fischer.
Goldschmidt, R. 1955. *Theoretical Genetics*. Berkeley: University of California Press, pp. 393–400.
Williams, G. C. 1957. Pleiotrophy, natural selection, and the evolution of senescence. *Evolution* 11:398–411.
Dobzhansky, T. 1959. Evolution of genes and genes in evolution. *Cold Spr. Harb. Symp. Quant. Biol.* 24:15–30.
Mayr, E. 1963. *Animal Species and Evolution*. Cambridge: Belknap Press of Harvard University Press, pp. 159, 264.
Cf. PATTERN EFFECT

PLETHORA THEORY [of metabolic rate]
Voit, E. 1901. Über die Grösse des Energiebedarfs der Tiere in

Hungerzustande. Z. *Biol.* 41:113–54.

Kleiber, M. 1961. *The Fire of Life.* New York: Wiley, p. 266.

POLAR EXPANSION THEORY [of cell cleavage]

Swann, M. M., & J. M. Mitchison. 1958. The mechanism of cleavage in animal cells. *Biol. Rev.* 33:103–35.

Rappaport, R. 1974. Cleavage. *Concepts of Development.* Edited by J. Lash & J. R. Whittaker. Stamford, Conn.: Sinauer Associates, pp. 76–98.

POLAR RELAXATION MECHANISM THEORY [cell cleavage]

Conklin, E. G. 1902. Karyokinesis and cytokinesis in the maturation, fertilization and cleavage of *Crepidula* and other gastropoda. *J. Acad. Nat. Sci. Phila.*, ser. 2, 12(1):1–21.

Wolpert, L. 1960. The mechanics and mechanism of cleavage. *Int. Rev. Cytol.* 10:163–216.

Rappaport, R. 1974. Cleavage. *Concepts of Development.* Edited by J. Lash and J. R. Whittaker. Stamford, Conn.: Sinauer Associates, pp. 76–98.

POLYCLIMAX *see* **CLIMAX**

POLYETHISM [division of labor among social insects]

Weir, J. 1958. Polyethism in workers of the ant *Myrmica. Insectes Sociaux* 5:97–128, 315–39.

Wilson, E. O. 1963. The social biology of ants. *Annu. Rev. Entomol.* 8:345–68.

POLYGENES [genetics]

Mather, K. 1941. Variation and selection of polygenic characters. *J. Genet.* 41:159–93.

Thoday, J. M. 1961. Location of polygenes. *Nature* 191:368–70.

______ . 1977. Effects of specific genes. *Proceedings of the International Conference on Quantitative Genetics*, Aug. 16–21, 1976. Edited by E. Pollack, O. Kempthorne, & T. B. Bailey. Ames: Iowa State University Press, pp. 141–59.

Cf. MULTIPLE FACTOR HYPOTHESIS

POLYGYNY HYPOTHESIS [evolution of territoriality]

McLaren, I. A. 1972. Polygyny as the adaptive function of breeding territory in birds. *Trans. Conn. Acad. Arts Sci.* 44:191–210.

Verner, J. 1977. On the adaptive significance of territoriality. *Amer. Nat.* 111:769–75.

POLYGYNY THRESHOLD *see* **ORIANS-VERNER MODEL**

POLYMERIC INHERITANCE *see* **MULTIPLE FACTOR HYPOTHESIS**

POLYPHYLESIS *see* **POLYPHYLETIC THEORY**

POLYPHYLETIC THEORY [of the origin of mammals]

Hopson, J. A., & A. W. Crompton. 1969. Origin of mammals. *Evol. Biol.* 3:15–72.

Cf. MONOPHYLY PRINCIPLE

POLYPHYLY PRINCIPLE [cladistic analysis]

Hennig, W. 1966. *Phylogenetic Systematics.* Urbana: University of Illinois Press, chap. 3.

Tuomikoski, R. 1967. Notes on some principles of phylogenetic systematics. *Ann. Entomol. Fenn.* 33:137–47.

Ashlock, P. D. 1971. Monophyly and associated terms. *Syst. Zool.* 20:63–69.

Nelson, G. J. 1971. Paraphyly and polyphyly: redefinitions. *Syst. Zool.* 20:471–72.

POPULATION [ecology; evolution; genetics]

Bakker, K. 1964. Backgrounds of controversies about population theories and their terminologies. *Z. Angew. Ent.* 53:187–208.

Jonckers, L. H. M. 1973. The concept of population in biology. *Acta Biotheor.* 22:78–108.

POPULATION CRASH *see* **OSCILLATIONS**

POPULATION FLUSH *see* **FOUNDER-FLUSH SPECIATION THEORY; OSCILLATIONS**

POPULATION THEORY [of honeybee recruitment]

Wenner, A. M. 1974. Information transfer in honeybees: a population approach. *Nonverbal Communication.* Edited by L. Krames, P. Pliner, & T. Alloway. New York: Plenum, vol. 1, pp. 133–69.

Gould, J. L. 1976. The dance-language controversy. *Q. Rev. Biol.* 51:211–44.

POPULATION VARIATION-ENVIRONMENTAL PATCHINESS HYPOTHESIS [ecology; evolution]

Grant, P. R. 1971. Variation in the tarsus length of birds in

 island and mainland regions. *Evolution* 25:599–614.
Cf. NICHE—VARIATION HYPOTHESIS

POPULATION WAVES [population genetics]
= waves of life

Chetverikov, S. S. 1905. Volny Zhizni [Waves of Life]. *Dnevnik Zootdeleniia* 3, no. 6.

Timofeef-Ressovsky, N. V. 1940. Mutations and geographical variation. *The New Systematics*. Edited by J. Huxley. London: Oxford University Press, pp. 73–136.

Adams, M. B. 1970. Towards a synthesis: population concepts in Russian evolutionary thought, 1925–1935. *J. Hist. Biol.* 3:107–29.

POSITION EFFECT [genetics]

Sturtevant, A. H. 1925. The effects of unequal crossing-over at the bar locus in *Drosophila*. *Genetics* 10:117–47.

Goldschmidt, R. B. 1946. Position effect and the theory of the corpuscular gene. *Experientia* 2:197–203, 250–56.

Lewis, E. B. 1950. The phenomenon of position effect. *Adv. Genet.* 3:73–115.

Goldschmidt, R. B. 1954. Different philosophies of genetics. *Science* 119:703–10.

Carlson, E. A. 1966. *The Gene; A Critical History*. Philadelphia: Saunders, chaps. 13–15.

Grant, V. 1971. *Plant Speciation*. New York: Columbia University Press, p. 128.

Cf. DUBININ EFFECT

PRAIRIE PENINSULA [ecology; biogeography]

Gleason, H. A. 1909. Some unsolved problems of the prairies. *Bull. Torr. Bot. Club* 36:265–71.

______ . 1923. The vegetational history of the middle west. *Ann. Assoc. Amer. Geogr.* 12:39–85.

Transeau, E. N. 1935. The prairie peninsula. *Ecology* 16: 423–37.

Benninghoff, W. S. 1963 (publ. 1964). The prairie peninsula as a filter barrier to postglacial plant migration. *Proc. Indiana Acad. Sci.* 73:116–24.

PRE-ADAPTATION HYPOTHESIS [ecology; biogeography]

Davenport, C. B. 1903. The animal ecology of Cold Spring Sand Spit, with remarks on the theory of adaptation. *Univ. Chicago Publ.* 10:1–22.

Cavalli-Svorza, L. L., & J. Lederberg. 1956. Isolation of preadaptive mutants in bacteria by sib-selection. *Genetics* 41:367–81.

Bennett, J. 1960. A comparison of selective methods and a test of the pre-adaptation hypothesis. *Heredity* 15:65–77.

Barr, T. C., Jr. 1968. Cave ecology and the evolution of Troglobites. *Evol. Biol.* 2:35–102.

Frazzetta, T. H. 1975. *Complex Adaptations in Evolving Populations.* Sunderland, Mass.: Sinauer Associates, chap. 8.

PRECOCITY THEORY [of meiosis]

Darlington, C. D. 1931. Cytological theory in relation to heredity. *Nature* 127:709–12.

———. 1931. Meiosis. *Biol. Rev.* 6:221–64.

PREDATION HYPOTHESIS [ecology]

Paine, R. T. 1966. Food web complexity and species diversity. *Amer. Nat.* 100:65–75.

———. 1971. A short-term experimental investigation of resource partitioning in a New Zealand rocky intertidal habitat. *Ecology* 52:1096–106.

Pianka, E. R. 1966. Latitudinal gradients in species diversity: a review of concepts. *Amer. Nat.* 100:33–46.

Menge, B. A., & J. P. Sutherland. 1976. Species diversity gradients: synthesis of the roles of predation, competition, and temporal heterogeneity. *Amer. Nat.* 110:351–69.

Houston, M. 1979. A general hypothesis of species diversity. *Amer. Nat.* 113:81–101.

Cf. BIOLOGICAL EXPLOITATION THEORY

PREDATION THEORY [bird flight origin] *see* **PREDATOR THEORY**

PREDATOR HYPOTHESIS [of avian clutch size]

Skutch, A. F. 1949. Do tropical birds rear as many young as they can nourish? *Ibis* 91:430–55.

Cody, M. L. 1966. A general theory of clutch size. *Evolution* 20:174–84.

Pianka, E. R. 1978. *Evolutionary Ecology.* 2d ed. New York: Harper & Row, p. 138.

PREDATOR MIMICRY HYPOTHESIS [evolution]

Zaret, T. M. 1977. Inhibition of cannibalism in *Cichla ocellaris* and hypothesis of predator mimicry among South American fishes. *Evolution* 31:421–37.

PREDATOR THEORY [of bird flight origin]

= cursorial predator theory = predation theory

Orstrom, J. H. 1974. *Archeopteryx* and the origin of flight. *Q. Rev. Biol.* 49:27–47.

PREDATOR-BUFFER HYPOTHESIS [evolution of territoriality]

Krebs, J.R. 1971. Territory and breeding density in the great tit, *Parus major* L. *Ecology* 52:2–22.

Verner, J. 1977. On the adaptive significance of territoriality. *Amer. Nat.* 111:769–75.

PREEXISTENCE THEORY [of heredity]

Bowler, P. 1971. Preformation and preexistence in the seventeenth century: a brief analysis. *J. Hist. Biol.* 4:221–44.

Farley, J. 1977. *The Spontaneous Generation Controversy from Descartes to Oparin*. Baltimore: Johns Hopkins Press, chap. 2.

PREFORMATION THEORY [of heredity]

Cole, F.J. 1930. *Early Theories of Sexual Generation*. Oxford: Clarendon Press, chap. 3.

Thompson, D'A.W. 1942. *On Growth and Form.* 2d ed. Cambridge: Cambridge University Press, pp. 82–89.

Bowler, P. 1971. Preformation and preexistence in the seventeenth century: a brief analysis. *J. Hist. Biol.* 4:221–44.

Cf. EPIGENESIS; DETERMINATION CONCEPT; TWO-LAYER THEORY

PREPATTERN-PRECURSOR THEORY [evolution]

Sondhi, K.C. 1963. The biological foundations of animal patterns. *Q. Rev. Biol.* 38:289–327.

Van Valen, L. 1970. An analysis of developmental fields. *Dev. Biol.* 23:456–77.

Frazzetta, T.H. 1975. *Complex Adaptations in Evolving Populations*. Sunderland, Mass.: Sinauer Associates, chap. 7.

PRESENCE-ABSENCE HYPOTHESIS [of the gene and gene mutations]

Bateson, W. 1903. The present state of knowledge of colour-heredity in mice and rats. *Proc. Zool. Soc. Lond.* 2:71–99.

______ . 1906. Presidential report: "progress of genetic research." *Proc. 3d Int. Congr. Genet.*, pp. 90–97.

Morgan, T.H., A.H. Sturtevant, H.J. Muller, & C.B. Bridges. 1923. *The Mechanism of Mendelian Heredity*. New York: Holt, pp. 216*ff*.

Swinburne, R. 1962. The presence and absence theory. *Ann. Sci.* 18:131–145.

Carlson, E.A. 1966. *The Gene: A Critical History*. Philadelphia: Saunders, pp. 58*ff*.

PRETRACHEATION THEORY [of insect wing origin]

Comstock, J.H., & J.G. Needham. 1898–99. The wings of insects. *Amer. Nat.* 32:43*ff*; 33:117*ff*.

Carpenter, F.M. 1966. The lower Permian insects of Kansas. II. The order Protorthoptera and Orthoptera. *Psyche* 73:46–88.

Kukalova-Peck, J. 1978. Origin and evolution of insect wings and their relation to metamorphosis, as documented by the fossil record. *J. Morph.* 156:53–126.

PREVAILING CLIMAX *see* **CLIMAX**

PRIMARY INTERGRADATION *see* **INTERGRADATION ZONES**

PRIMORDIAL SOUP or BROTH [chemical origin of life on earth]

= prebiotic soup = probiotic soup = primitive soup = primeval broth = primeval soup = nutrient soup = primeval nutritive soup = Miller-Urey soup = thick soup = organic milieu = Haldane soup

Oparin, A.I. 1938. *The Origin of Life*. New York: Macmillan, chaps. 5–6.

Haldane, J. B. S. 1945. A new theory of the past. *Amer. Sci.* 33:129–45.

Bernal, J. D. 1951. *The Physical Basis of Life*. London: Routledge & Kegan Paul.

Urey, H. C. 1952. On the early chemical history of the earth and the origin of life. *Proc. Natl. Acad. Sci.* 38:351–63.

Miller, S. L. 1953. A production of amino acids under possible primitive earth conditions. *Science* 117:528–29.

Fox, S. W., ed. 1965. *The Origins of Prebiological Systems and of Their Molecular Matrices*. New York: Academic Press.

PRIOR-DIVERSIFICATION THEORY *see* **UPLAND THEORY** [of angiosperm origin]

PROCHRONISM, LAW OF

= Gosse's law

Gosse, P. H. 1857. *Omphalos—An Attempt to Untie the Geological Knot*. London: Van Voorst.

Bateson, G. 1970. On empty-headedness among biologists and state boards of education. *Bioscience* 20:819.

PRODUCTIVITY [ecological]

Macfadyen, A. 1948. The meaning of productivity in biological systems. *J. Anim. Ecol.* 17:75–80.

———. 1963. *Animal Ecology: Aims and Methods.* 2d ed. London: Pitman, chap. 11.

Engelmann, M. D. 1966. Energetics, terrestrial field studies, and animal productivity. *Adv. Ecol. Res.* 3:73–115.

PRODUCTIVITY HYPOTHESIS [of species diversity]

Connell, J. H., & E. Orias. 1964. The ecological regulation of species diversity. *Amer. Nat.* 98:399–441.

Pianka, E. R. 1966. Latitudinal gradients in species diversity: a review of concepts. *Amer. Nat.* 100:33–46.

Margalef, R. 1969. Diversity and stability: a practical proposal and a model of interdependence. *Brookh. Symp. Biol.* 22:25–37.

Pielou, E. C. 1975. *Ecological Diversity.* New York: Wiley, pp. 129*ff.*

Huston, M. 1979. A general hypothesis of species diversity. *Amer. Nat.* 113:81–101.

PROGRESS [evolution]

Simpson, G. G. 1949. *The Meaning of Evolution.* New Haven, Conn.: Yale University Press, pp. 239*ff.*

Thoday, J. M. 1958. Natural selection and biological progress. *A Century of Darwin.* Edited by S. A. Barnett. Cambridge: Harvard University Press, pp. 313–33.

Ayala, F. J. 1974. The concept of biological progress. *Studies in the Philosophy of Biology.* Edited by F. J. Ayala & T. Dobzhansky. Berkeley: University of California Press, pp. 339–55.

PROGRESSION RULE [taxonomy]

= Hennig's progression rule

Hennig, W. 1966. The Diptera fauna of New Zealand as a problem in systematics and zoogeography. *Pac. Insects Monogr.*, vol. 9.

Ashlock, P. D. 1974. The uses of cladistics. *Annu. Rev. Ecol. Syst.* 5:81–99.

PROGRESSIONISM [19th-century evolution]

Miller, H. 1850. *Footprints of the Creator.* Boston: Gould, Kendall & Lincoln.

Mayr, E. 1976. *Evolution and the Diversity of Life.* Cambridge: Belknap Press of Harvard University Press, p. 285.

PROGRESSIVE DEVIATION, PRINCIPLE OF *see* **VON BAER'S LAWS**

PROGRESSIVE SELECTION *see* **DIRECTIONAL SELECTION**

PROTEIN HYPOTHESIS [of primate feeding behavior]
Hamilton, W. J., & C. D. Busse. 1978. Primate carnivory and its significance to human diets. *Bioscience* 28:761–66.

PROTEIN STRUCTURE THEORIES [historical]
Fruton, J. S. 1979. Early theories of protein structure. *Ann. N.Y. Acad. Sci.* 325:1–18.

PROTEINS-FIRST HYPOTHESIS [origin of life]
Moody, A R. 1970. *Introduction to Evolution.* 3d ed. New York: Harper & Row, pp. 117*ff.*
Fox, S. W., & K. Dose. 1972. *Molecular Evolution and the Origin of Life.* San Francisco: Freeman, p. 239.

PROTOPLASM THEORY [of life]
Dujardin, F. 1835. Recherches sur les organismes inférieurs. *Ann. Sci. Nat.*, ser. 2, 4:343–77.
Geison, G. 1969. The protoplasmic theory of life and the vitalist-mechanist debate. *Isis* 60:273–92.
Kohmler, R. E. 1972. The reception of Eduard Buchner's discovery of cell-free fermentation. *J. Hist. Biol.* 5:327–53.
_____ . 1973. The enzyme theory and the origins of biochemistry. *Isis* 64:181–96.
Cf. ENZYME THEORY; ZYMASE HYPOTHESIS

PROTOPLASMIC BRIDGE THEORY [of nerve fiber development]
= Hensen's theory = PLASMODESM THEORY
Hensen, V. 1864. Über die Entwickelung des Gewebes und der Nerven im Schwanze der Froschlarve. *Virchows Arch. Path. Anat.* 31:51–73.
_____ . 1861. Ueber die Nerven im Schwanz der Froschlarven. *Arch. Mikr. Anat.* 4:111–24.
Billings, S. M. 1971. Concepts of nerve fiber development, 1839–1930. *J. Hist. Biol.* 4:275–305.
Cf. MULTICELLULAR THEORY; NEURON THEORY; OUTGROWTH THEORY

PROTOSTOMIA-DEUTEROSTOMIA THEORY [vertebrate evolution]

Bateson, W. 1886. The ancestry of the Chordata. *Q. J. Microsc. Sci.* 26:535–71.

Kerkut, G. A. 1960. *The Implications of Evolution*. New York: Pergamon.

Løvtrup, S. 1975. Validity of the Protostomia-Deuterostomia theory. *Syst. Zool.* 24:96–108.

PROTOVIRUS HYPOTHESIS [role of viruses in cancer]

Temin, H. M. 1970. Malignant transformation of cells by viruses. *Perspect. Biol. Med.* 14:11–26.

———. 1971. The protovirus hypothesis. *J. Natl. Cancer Inst.* 46(2):iii–viii.

———. 1972. The protovirus hypothesis and cancer. *RNA Viruses and Host Genomes in Oncogenesis*. Edited by P. Emmelot & P. Bentvelzen. Amsterdam & London: North Holland, pp. 351–63.

Gross, L. 1974. Facts and theories on viruses causing cancer and leukemia. *Proc. Natl. Acad. Sci.* 71:2013–17.

Cf. ONCOGENE HYPOTHESIS

PROVIRUS HYPOTHESIS [role of viruses in cancer]

Temin, H. M. 1964. Nature of the provirus of Rous sarcoma. *Natl. Cancer Inst. Monogr.* 17:557–70.

———. 1971. The protovirus hypothesis. *J. Natl. Cancer Inst.* 46(2):iii–viii.

Cf. PROTOVIRUS HYPOTHESIS

PROXIMATE FACTORS *see* **ULTIMATE AND PROXIMATE FACTORS**

PRUDENT-PREDATOR CONCEPT [evolution; feeding behavior]

Slobodkin, L. B. 1961. *Growth and Regulation of Animal Populations*. New York: Holt, Rinehart & Winston, chap. 13.

———. 1968. How to be a predator. *Amer. Zool.* 8:43–51.

———. 1974. Prudent predation does not require group selection. *Amer. Nat.* 108:665–78.

Maynard Smith, J., & M. Slatkin. 1973. The stability of predator-prey systems. *Ecology* 54:384–91.

Maiorana, V. C. 1976. Reproductive value, prudent predators, and group selection. *Amer. Nat.* 110:486–89.

Mertz, D. B., & M. J. Wade. 1976. The prudent prey and the prudent predator. *Amer. Nat.* 110:489–96.

PRZIBRAM'S LAW [insect growth rate]

Przibram, H., & F. Megusar. 1912. Wachstummessungen an *Sphrodomantis*. *Arch. Entw. Mech.* 34:680–741.

Przibram, H. 1930. *Connecting Laws in Animal Morphology*. London: University of London Press.

Thompson, D'A. W. 1942. *On Growth and Form*. 2d ed. Cambridge: Cambridge University Press, pp. 165, 165n.

Cf. DYAR'S LAW

PSEUDANTHIUM HYPOTHESES *see* **MULTIAXIAL FLOWER THEOREM**

PSEUDOALLELISM [genetics]

McClintock, B. 1944. The relation of homozygous deficiencies to mutations and allelic series in maize. *Genetics* 29:478–502.

Lewis, E. B. 1951. Pseudoallelism and gene evolution. *Cold Spr. Harb. Symp. Quant. Biol.* 16:159–74.

______. 1955. Some aspects of position pseudoallelism. *Amer. Nat.* 89:73–89.

______ . 1967. Gene and gene complexes. *Heritage from Mendel*. Edited by R. A. Brink & E. D. Styles. Madison: University of Wisconsin Press, pp. 17–47.

PUMP-LEAK HYPOTHESIS [biophysics]

Tosteson, D. C. 1964. Regulation of cell volume by sodium and potassium transport. *The Cellular Functions of Membrane Transport*. Edited by J. F. Hoffman. Englewood Cliffs, N.J.: Prentice-Hall, pp. 3–22.

Orloff, J., & J. S. Handler. 1964. Mechanism of action of antidiuretic hormones on epithelial structures. *The Cellular Functions of Membrane Transport*. Edited by J. F. Hoffman. Englewood Cliffs, N.J.: Prentice-Hall, pp. 251-68.

PUNCTUATED EQUILIBRIA THEORY [evolution]

Eldredge, N., & S. J. Gould. 1972. Punctuated equilibria: an alternative to phyletic gradualism. *Models in Paleobiology*. Edited by T. J. M. Schopf. San Francisco: Freeman, Cooper, pp. 82–115.

Hecht, M. K., N. Eldredge, & S. J. Gould. 1974. Morphological transformation, the fossil record, and the mechanisms of evolution: a debate. *Evol. Biol.* 7:295–308.

Gould, S. J., & N. Eldredge. 1977. Punctuated equilibria: the tempo and mode of evolution reconsidered. *Paleobiology* 3:115–51.

PUNNETT'S THEORY OF MIMICRY *see* **MIMICRY; MIMETIC POLYMORPHISM**

PURE LINE THEORY [genetics]

Johannsen, W. L. 1903. *Ueber Erblichkeit in Populationen und in reinen Linien.* Jena: Fischer. (1955. Concerning heredity in populations and in pure lines. *Selected Readings in Biology for Natural Sciences.* vol. 3. Transl. by H. Gall & E. Putschar. Chicago: University of Chicago Press, 172-215.)

Castle, W. E. 1914. Pure lines and selection. *J. Hered.* 5:93–97.

Provine, W. B. 1971. *The Origins of Theoretical Population Genetics.* Chicago: University of Chicago Press, chap. 4.

Wanscher, J. H. 1975. The history of Wilhelm Johanssen's genetical terms and concepts from the period 1903 to 1926. *Centaurus* 19:125–47.

Spiess, E. B. 1977. *Genes in Populations.* New York: Wiley, chap. 6.

PURIFYING SELECTION *see* **HALDANE'S DILEMMA**

PÜTTER'S THEORY [aquatic animal feeding behavior]

Pütter, A. 1908. Die Ernährung der Wassertiere. Z. *Allg. Physiol.* 7:288–320.

Allee, W. C., A. E. Emerson, O. Park, T. Park, & K. P. Schmidt. 1949. *Principles of Animal Ecology.* Philadelphia: Saunders, p. 444.

Jørgensen, C. B. 1976. August Pütter, August Krogh, and modern ideas on the use of dissolved organic matter in aquatic environments. *Biol. Rev.* 51:291–328.

PYRAMID OF NUMBERS [ecology]

Elton, C. S. 1935. *Animal Ecology.* 2d ed. London: Sidgwick & Jackson, chap. 5.

______. 1966. *The Pattern of Animal Communities.* London: Methuen; New York: Wiley, pp. 189, 375.

Kormondy, E. J. 1969. *Concepts of Ecology.* Englewood Cliffs, N.J.: Prentice-Hall, pp. 31–33.

[Q]

QUANTAL MITOSIS CONCEPT [cytology; morphogenesis]

Konigsberg, I. R., & P. A. Buckley. 1974. Regulation of the cell cycle and myogenesis by cell-medium interaction. *Concepts of Development*. Edited by J. Lash & J. R. Whittaker. Stamford, Conn.: Sinauer Associates, pp. 179–93.

QUANTUM EVOLUTION

Simpson, G. G. 1944. *Tempo and Mode in Evolution*. New York: Columbia University Press, pp. 206*ff*.

——— . 1953. *The Major Features of Evolution*. New York: Columbia University Press, pp. 389*ff*.

Dodson, M. M. 1975. Quantum evolution and the fold catastrophe. *Evol. Theory* 1:107–18.

Cf. CATASTROPHE THEORY; PHYLETIC EVOLUTION; QUANTUM SPECIATION; SALTATION

QUANTUM SPECIATION

= SALTATION = saltational speciation

Simpson, G. G. 1944. *Tempo and Mode in Evolution*. New York: Columbia University Press, p. 206.

Grant, V. 1963. *The Origin of Adaptations*. New York: Columbia University Press, pp. 456*ff*.

———. 1971. *Plant Speciation*. New York: Columbia University Press, pp. 114*ff*.

Carson, H. L. 1975. The genetics of speciation at the diploid level. *Amer. Nat.* 109:83–92.

Cf. QUANTUM EVOLUTION

QUETELET'S LAW [statistics]

Quetelet, L. A. J. 1871. *Anthropométrie ou Mesure des différentes facultés de l'Homme*. Bruxelles: Muquardt.

Roemer, Th. 1936. Die Bedeutung des Gesetzes der Parallelvariation für die Pflanzenzüchtung. *Nova Acta Leopold.* 4:351–65.

Thompson, D'A. W. 1942. *On Growth and Form*. 2d ed. Cambridge: Cambridge University Press, p. 122.

Hutchinson, G. E. 1978. *An Introduction to Population Ecology*. New Haven, Conn.: Yale University Press, p. 18.

[R]

r AND K SELECTION [evolution; population genetics]

MacArthur, R. H., & E. O. Wilson. 1967. *The Theory of Island Biogeography*. Princeton, N.J.: Princeton University Press.

Hairston, N. G., D. W. Tinkle, & H. M. Wilbur. 1970. Natural selection and the parameters of population growth. *J. Wildl. Manage.* 34:681–90.

Pianka, E. R. 1970. On r and K selection. *Amer. Nat.* 104:592-97.

______ . 1972. r and K selection or b and d selection? *Amer. Nat.* 106:581–88.

Roughgarden, J. 1971. Density dependent natural selection. *Ecology* 52:453–68.

Gadgil, M., & O. T. Solbrig. 1972. The concept of r and K selection: evidence from wild flowers and some theoretical considerations. *Amer. Nat.* 106:14–31.

McNaughton, S. J. 1975. r- and K-selection in *Typha. Amer. Nat.* 109:251–61.

Nichols, J. D., W. Conley, B. Batt, & A. R. Tipton. 1976. Temporally dynamic reproductive strategies and the concept of r- and K-selection. *Amer. Nat.* 110:995–1005.

RANDOM DRIFT *see* **RANDOM WALK**

RANDOM ENVIRONMENT CONCEPT [ecology]

Lewontin, R. C., & D. Cohen. 1969. On population growth in randomly varying environments. *Proc. Natl. Acad. Sci.* 62:1056–60.

Smith, W. L., & W. E. Wilkinson. 1969. On branching processes in random environments. *Ann. Math. Statis.* 40:814–27.

Turelli, M. 1977. Random environments and stochastic calculus. *Theor. Pop. Biol.* 12:140–78.

RANDOM GENETIC DRIFT *see* **GENETIC DRIFT**

RANDOM WALK [gene frequencies]

= random drift

Ayala, F.J. 1974. Biological evolution: Natural selection or random walk? *Amer. Sci.* 62:692–701.

Spiess, E. B. 1977. *Genes in Populations.* New York: Wiley, pp. 338*ff*.

Cf. NEUTRAL THEORY [of genetic fitness]; NON-DARWINIAN EVOLUTION

RAUNKIAER'S LAW OF FREQUENCY *see* **FREQUENCY, LAW OF; FREQUENCY-DISTRIBUTION CURVES**

REACTION TYPE *see* **PHENOTYPE**

REACTIVE SITE [biophysics]
= active center

Christensen, H. N. 1970. Concept of the reactive site in biological transport. *Physical Principles of Biological Membranes*. Edited by F. Snell, J. W. Wolken, G. Iverson, & J. Lam. New York: Gordon & Breach, pp. 397–413.

REALITY, THEORY OF [in classification; microbiology]

Kaufmann, F. 1971. Eine neue, realistische Klassifikation. *Zbl. Bakter. I. Orig.* 217:198–201.

______ . 1971. Eine neue Vereinfachte Nomenklatur. *Zbl. Bakter. I. Orig.* 217:202–05.

______ . 1973. On the realistic classification and evaluation of serology. *Acta Path. Microbiol. Scand. B* 81:198–202.

RECAPITULATION, LAW OF or THEORY [evolution]
= BIOGENESIS

Hurst, C. H. 1893. Biological theories. III. The recapitulation theory. *Nat. Sci.* 2:195–200.

Conklin, E. G. 1928. Embryology and evolution. *Creation by Evolution*. Edited by F. Mason. New York: Macmillan, pp. 62–80.

de Beer, G. 1940. *Embryos and Ancestors*. Oxford: Clarendon Press.

Stebbins, G. L. 1950. *Variation and Evolution in Plants*. New York: Columbia University Press, p. 488.

Sporne, K. R. 1956. The phylogenetic classification of the angiosperms. *Biol. Rev.* 31:1–29.

Thorne, R. F. 1958. Some guiding principles of angiosperm phylogeny. *Brittonia* 10:72–77.

Gould, S. J. 1977. *Ontogeny and Phylogeny*. Cambridge: Harvard University Press, pp. 13*ff*.

Løvtrup, S. 1978. On von Baerian and Haeckelian recapitulation. *Syst. Zool.* 27:348–52.

RECEPTOR THEORY *see* **EHRLICH'S SIDE-CHAIN THEORY**

RECIPROCAL ALTRUISM *see* **ALTRUISM**

RECOMBINATION *see* **GENE CONVERSION**

RECOMBINATION STRATEGY-VARIATION HYPOTHESIS [of population diversity]

Mather, K. 1953. The genetical structure of populations. *Symp. Soc. Exp. Biol.* 7:66–95.

Zangerl, A. R., S. T. A. Pickett, & F. A. Bazzaz. 1977. Some hypotheses on variation in plant populations and an experimental approach. *Biologist* 59:113–22.

RECOMBINATIONAL SPECIATION

Grant, V. 1971. *Plant Speciation*. New York: Columbia University Press, p. 193 & chap. 14.

RECTANGULAR MODEL OF PHYLOGENY [evolution]

Eldredge, N. 1971. The allopatric model and phylogeny in paleozoic invertebrates. *Evolution* 25:156–67.

Stanley, S. M. 1975. Clades versus clones in evolution: why we have sex. *Science* 190:382–83.

______ . 1975. A theory of evolution above the species level. *Proc. Natl. Acad. Sci.* 72:646–50.

RED QUEEN HYPOTHESIS [evolution]

= constant extinction, law of

Van Valen, L. 1973. A new evolutionary law. *Evol. Theory* 1:1–30.

______ . 1976. Energy and evolution. *Evol. Theory* 1:179–229.

______ . 1977. The red queen. *Amer. Nat.* 111:809–10.

Foin, T. C., J. W. Valentine, & F. J. Ayala. 1975. Extinction of taxa and Van Valen's law. *Nature* 257:514–15.

Darlington, P. J. 1977. The cost of evolution and the imprecision of adaptation. *Proc. Natl. Acad. Sci.* 74:1647–51.

Maynard Smith, J. 1976. A comment on the red queen. *Amer. Nat.* 110:325–30.

Castrodeza, C. 1979. Non-progressive evolution, the red queen hypothesis, and the balance of nature. *Acta Biotheor.* 28:11–18.

REDOX HYPOTHESIS [of active transport]

= redox pump theory

Lundergårdh, H. 1955. Mechanisms of absorption, transport, accumulation, and secretion of ions. *Annu. Rev. Plant Physiol. 6:1–24.*

Robertson, R. N. 1960. Ion transport and respiration. *Biol. Rev.* 35:231–64.

REDUCTION OF POLYMORPHY, LAW OF *see* **HAGEDOORN EFFECT**

REDUCTION SERIES [plant evolution]

Woodson, R.E. 1935. Observations on the inflorescences of the *Apocynaceae*. *Ann. Mo. Bot. Gard.* 22:1–48.

Iltis, H. H. 1957. Studies in the *Capparidaceae*. III. Evolution and phylogeny of the western North American Cleomoideae. *Ann. Mo. Bot. Gard.* 44:77–119.

Asama, K. 1960. Evolution of the leaf forms through the ages explained by the successive retardation and neoteny. *Tohuku Univ. Science Reports, 2d ser. Geology, spec. v. 4,* pp. 252–80.

———. 1974. Origin of angiosperms inferred from the evolution of leaf forms. *Birbal Sahni Inst. Paleobotany Special Publ.* no. 1, pp. 1–4.

———. 1975. *Evolutionary Biology in Plants. IV. The Origin of the Angiosperms.* Tokyo: Sanseido.

REFLEXIVE SELECTION [of prey by predator]

Moment, G. B. 1962. Reflexive selection: a possible answer to an old puzzle. *Science* 136:262-63.

REFUGIUM THEORY [biogeography]

= survival theory

Hoppe, G. 1963. Some comments on the "ice-free refugia" of northwestern Scandinavia. *North Atlantic Biota and Their History.* Edited by A. Löve & D. Löve. New York: Macmillan, pp. 321–35.

Lindroth, C. H. 1970. Survival of animals and plants on ice-free refugia during the Pleistocene glaciations. *Endeavour* 29:129–34.

Vailleumier, B. 1971. Pleistocene changes in the fauna and flora of South America. *Science* 173:771–80.

Simpson, B. B., & J. Hoffer. 1978. Speciation patterns in the Amazon forest biota. *Annu. Rev. Ecol. Syst.* 9:497–518.

Cf. NUNATAK HYPOTHESIS

REGULATORY HYPOTHESIS [immunology; evolution]

Wilson, A. C., L. R. Maxson, & V. M. Sarich. 1974. Two types of molecular evolution. Evidence from studies of interspecific hybridization. *Proc. Natl. Acad. Sci.* 71:2843–47.

Wilson, A. C., V. M. Sarich, & L. R. Maxson. 1974. The importance of gene rearrangement in evolution: evidence from studies on rates of chromosomal, protein, and anatomical evolution. *Proc. Natl. Acad. Sci.* 71:3028–30.

Cf. IMMUNOLOGICAL INERTIA

REJECTER SPECIES [bird behavior]

Rothstein, S. I. 1975. Evolutionary rates and host defenses against avian brood parasitism. *Amer. Nat.* 109:161–76.

———. 1975. An experimental and teleonomic investigation

of avian brood parasitism. *Condor* 77:250–71.
Cf. ACCEPTER SPECIES

REJUVENESCENCE [in paramecia; development]

Jennings, H. S. 1929. Genetics of the Protozoa. *Bibliogr. Genet.* 5:105–330.

Sonneborn, T. M. 1954. The relation of autogamy to senescence and rejuvenescence in *Paramecium aurelia. J. Protozool.* 1:38–53.

RELATED DEVIATIONS, LAW OF *see* **KRENKE'S RULE**

RELAY EFFECT [evolution]

Simpson, G. G. 1964. *This View of Life*. New York: Harcourt, Brace & World.

Dobzhansky, T. 1972. Darwinian evolution and the problem of extraterrestrial life. *Perspect. Biol. Med.* 15:157–75.

RELEVANCE [in taxonomy]

Cain, A. J., & G. A. Harrison. 1958. An analysis of the taxonomist's judgment of affinity. *Proc. Zool. Soc. Lond.* 131:85–98.

Crovello, T. J. 1968. Different concepts of relevance in a numerical taxonomic study. *Nature* 218:492.

Baum, B. R. 1973. The concept of relevance in taxonomy with special emphasis on automatic classification. *Taxon* 22:329–32.

RELICT [ecology]

Clements, F. E. 1934. The relict method in dynamic ecology. *J. Ecol.* 22:39–68.

RELICT HYPOTHESIS [of bipolar distribution of animals]

Theel, H. 1911. Priapulids and sipunculids dredged by the Swedish Antarctic expedition 1901–1903 and the phenomenon of bipolarity. *K. Sven. Vetenskapsakad. Handl.* 47(1): 1–36.

Allee, W. C., & K. P. Schmidt. 1951. *Ecological Animal Geography*. 2d ed., rev. by R. Hesse. New York: Wiley, pp. 333*ff*.

RELICT SPECIES [biogeography; evolution]

Fryxell, P. A. 1962. The "relict species" concept. *Acta Biotheor.* 15:105–18.

RENSCH'S RULE [relative growth of wing tips and feathers] = WING RULE

Allee, W. C., & K. P. Schmidt. 1951. *Ecological Animal Geography*. 2d ed., rev. by R. Hesse. New York: Wiley, p. 464.

Rensch, B. 1960. *Evolution Above the Species Level*. New York: Columbia University Press, p. 43.

Cf. ALLEN'S RULE; BERGMANN'S RULE; GLOGER'S RULE

REPAIR HYPOTHESIS [of the origin of complete mutations]

Freese, E., & E. B. Freese. 1966. Production of pure mutant clones by repair of inactivating DNA alterations in phage T4. *Genetics* 54:1055–67.

Nasim, A., & C. Auerbach. 1967. The origin of complete and mosaic mutants from mutagenic treatment of single cells. *Mutat. Res.* 4:1–14.

Dubinin, N. P., & V. N. Soyfer. 1969. Chromosome breakage and complete genic mutation production in molecular terms. *Mutat. Res.* 8:353–65.

Auerbach, C., & B. J. Kilbey. 1971. Mutation in eukaryotes. *Annu. Rev. Genet.* 5:163–218.

Cf. DUAL-MECHANISM HYPOTHESIS; LETHAL-HIT HYPOTHESIS; MASTER-STRAND HYPOTHESIS

REPRESSOR-OPERON THEORY *see* **OPERON THEORY**

REPRODUCTIVE ASSURANCE HYPOTHESIS [evolution; biogeography]

Jain, S. K. 1976. The evolution of inbreeding in plants. *Annu. Rev. Ecol. Syst.* 7:469–95.

Cf. BAKER'S LAW

REPRODUCTIVE COMMUNITY *see* **MENDELIAN POPULATION**

REPRODUCTIVE EFFORT [evolution]

Tinkle, D. W. 1969. The concept of reproductive effort and its relation to the evolution of life histories of lizards. *Amer. Nat.* 103:501–16.

Goodman, D. 1974. Natural selection and a cost ceiling on reproductive effort. *Amer. Nat.* 108:247–68.

Hirshfield, M.F., & D. W. Tinkle. 1975. Natural selection and the evolution of reproductive effort. *Proc. Natl. Acad. Sci.* 72:2227–31.

Stearn, S. C. 1976. Life-history tactics: a review of the ideas. *Q. Rev. Biol.* 51:3–47.

REPRODUCTIVE ISOLATION [evolution]

Mayr, E. 1959. Isolation as an evolutionary factor. *Proc. Amer. Philos. Soc.* 103:221–30.

______ . 1963. *Animal Species and Evolution*. Cambridge: Belknap Press of Harvard University Press, chap. 5.

Dobzhansky, T. 1970. *Genetics of the Evolutionary Process*. New York: Columbia University Press, chaps. 10-11.

Ehrman, L. 1971. Natural selection for the origin of reproductive isolation. *Amer. Nat.* 105:479–83.

Soans, A. B., D. Pimentel, & J. S. Soans. 1974. Evolution of reproductive isolation in allopatric and sympatric populations. *Amer. Nat.* 108:117–24.

Cf. GEOGRAPHIC ISOLATION; ISOLATING MECHANISMS

REPRODUCTIVE VALUE [evolution]

Fisher, R. A. 1930. *The Genetical Theory of Natural Selection*. Oxford: Clarendon Press, chap. 2.

Williams, G. C. 1966. Natural selection, the costs of reproduction, and a refinement of Lack's principle. *Amer. Nat.* 100:687–90.

REPULSION THEORY [of phyllotaxy; morphogenesis]

Schmucker, T. 1933. Zur Entwicklungsphysiologie der schraubigen Blattstellung. *Planta* 19:139–53.

Snow, M., & R. Snow. 1948. On the determination of leaves. *Symp. Soc. Exp. Biol.* 2:263–75.

Wardlaw, C. W. 1952. *Phylogeny and Morphogenesis*. London: Macmillan, p. 395.

RESILIENCE [of ecosystems]

Holling, C. S. 1973. Resilience and stability of ecological systems. *Annu. Rev. Ecol. Syst.* 4:1–23.

Westman, W. W. 1978. Measuring the inertia and resilience of ecosystems. *Bioscience* 28:705–10.

RESOURCE CONCENTRATION HYPOTHESIS [ecology]

Root, R. B. 1973. Organization of a plant-arthropod association in simple and diverse habitats: the fauna of collards (*Brassica oleracea*). *Ecol. Monogr.* 43:95–124.

RESOURCE HARVESTING *see* **OPTIMAL FORAGING THEORY**

RESOURCE PARTITIONING [ecology]

Schoener, T. W. 1974. Resource partitioning in ecological communities. *Science* 185:27–39.

Cf. HABITAT SHIFT; NICHE SHIFT

RESOURCE-PREDICTABILITY VARIATION HYPOTHESIS
[of the niche; ecology]

Valentine, J. W. 1971. Resource supply and species diversity patterns. *Lethaia* 4:51–61.

Ayala, F. J., & J. W. Valentine. 1974. Genetic variability in the cosmopolitan deep-water ophiuram, *Ophiomusium lymani*. *Mar. Biol.* 27:51–57.

Soule, M. 1976. Allozyme variation: its determinants in space and time. *Molecular Evolution*. Edited by F. J. Ayala. Sunderland, Mass.: Sinauer Associates, pp. 60–77.

RETICULAR THEORY [of intercellular communication]
= continutity theory = nerve-net theory

Cajal, S. Ramon y. 1954. *Neuron Theory or Reticular Theory?* Transl. by M. O. Purkiss & C. A. Fox. Madrid: Consejo Superior de Investigationes cientificas Instituto "Ramon y Cajal."

French, R. D. 1970. Some concepts of nerve structure and function in Britain, 1875–1885: background to Sir Charles Sherrington and the synapse concept. *Med. Hist.* 14:154–65.

Grundfest, H. 1975. History of the synapse as a morphological and functional structure. *Golgi Centennial Symposium*. Edited by M. Santini. New York: Raven, pp. 39–50.

REVERSION *see* **GENE CONVERSION**

ROBERTSONIAN CHANGES [fusion and fission of chromosomes]
= Robertson's rule

Robertson, W. R. B. 1916. Chromosome studies. I. Taxonomic relationships shown in the chromosomes of Tettigidae and Acrididae. V-shaped chromosomes and their significance in Acrididae, Locustidae and Gryllidae: chromosomes and variation. *J. Morph.* 27:179–331.

Matthey, R. 1939. La loi de Robertson et la formule chromosomiale chez deux Lacertiens: *Lacerta ocellata* Daud., *Psammodromus hispanicus* Fitz. *Cytologia* 10:32–39.

Mayr, E. 1963. *Animal Species and Evolution*. Cambridge: Belknap Press of Harvard University Press, chap. 15.

White, M. J. D. 1969. Chromosomal rearrangements and speciation in animals. *Annu. Rev. Genet.* 3:75–98.

———. 1973. *Animal Cytology and Evolution*. 3d ed. Cambridge: Cambridge University Press, pp. 229–30.

Ohno, S. 1970. *Evolution by Gene Duplication*. New York &

Berlin: Springer-Verlag, chap. 7.

Dobzhansky, T., F. J. Ayala, G. L. Stebbins, & J. W. Valentine. 1977. *Evolution.* San Francisco: Freeman, p. 92.

ROOT EFFECT [physiology]

Root, R. W., L. Irving, & E. C. Black. 1939. The effect of hemolysis upon the combination of oxygen with the blood of some marine fishes. *J. Cell. Comp. Physiol.* 13:303–13.

Schmidt-Nielsen, K. 1979. *Animal Physiology: Adaptation and Environment.* Cambridge: Cambridge University Press, pp. 70, 430–32.

ROSA'S RULE [of progressive reduction of variability]

Fechner, G. T. 1873. *Einige Ideen zur Schöpfungs- und Entwicklungsgeschichte der Organismen.* Leipzig: Breitkopf & Härtel.

Rosa, D. 1899. *La Riduzione Progressiva della Variabilità e i suoi Rapporti Coll'estinzione e Coll'origine della Specie.* Torino: Clausen.

Hennig, W. 1966. *Phylogenetic Systematics.* Urbana: University of Illinois Press, pp. 218–20.

ROSA'S THEORY OF HOLOGENESIS *see* **HOLOGENESIS**

ROUX-WEISMANN THEORY *see* **GERMPLASM THEORY**

RPS (REPEATING POLYMORPHIC SET)

Mayen, S. V. 1973. Plant morphology in its nomothetical aspects. *Bot. Rev.* 39:205–60.

Cf. HOMOLOGOUS SERIES, LAW OF

RUBNER'S HYPOTHESIS [longevity; living intensity]

Rubner, M. 1908. *Das problem der Lebensdauer und Seine Beziehungen zu Wachstum und Ernahrung.* Munich & Berlin: Oldenbourg.

Northrop, J. H. 1926. Carbon dioxide production and duration of life of *Drosophila* cultures. *J. Gen. Physiol.* 9:319–24.

Pearl, R. 1927. *The Rate of Living.* New York: Knopf.

RUBNER'S LAW OF COMPENSATION [temperature regulation in birds]

= compensation theory

Rubner, M. 1910. Über Kompensation und Summation von funktionellen Leistungen des Körpers. *Sber. K. Preuss. Akad. Wiss.* 16:316–24.

Kleiber, M. 1961. *The Fire of Life*. New York: Wiley, chap. 15.

Kendeigh, S. C. 1969. Energy responses of birds to their thermal environments. *Wilson Bull.* 81:441–49.

RUBNER'S SPECIFIC DYNAMIC EFFECT *see* **SPECIFIC DYNAMIC EFFECT**

RUDERAL STRATEGY [in plants]

Grime, J. P. 1977. Evidence for the existence of three primary strategies in plants and its relevance to ecological and evolutionary theory. *Amer. Nat.* 111:1169–94.

RULE OF DEVELOPMENT *see* **FAHRENHOLZ'S RULE**

[S]

SACH'S RULE [cell division in plants]

= HERTWIG'S RULE [in animals]

Schwendener, S. 1860. Ueber den Bau und das Wachsthum des Flechtenthallus. *Viert. Naturf. Ges. Zürich,* 1860:272–96.

Sachs, J. von. 1877. Über die Anordnung der Zellen in jüngsten Pflanzentheilen. *Verh. Phys.-Med. Ges. Würzburg* 11:219–42.

Wilson, E. B. 1925. *The Cell in Development and Heredity*. 3d ed. New York: Macmillan, p. 982.

Thompson, D'A. W. 1942. *On Growth and Form*. 2d ed. Cambridge: Cambridge University Press, p. 481.

Bonner, J. T. 1965. *Size and Cycle*. Princeton, N.J.: Princeton University Press, p. 36.

Løvtrup, S. 1974. *Epigenetics*. New York: Wiley, pp. 128–29.

SALTATION [evolution]

= QUANTUM SPECIATION = saltatory speciation = saltational speciation

Mivart, St. G. 1871. *Genesis of Species*. London: Macmillan.

Lewis, H. 1966. Speciation in flowering plants. *Science* 152:167–72.

Van Steenis, C. G. G. J. 1969. Plant speciation in Malesia, with special reference to the theory of non-adaptive saltatory evolution. *Biol. J. Linn. Soc.* 1:97–133.

Cf. KRENKE'S RULE; PHYLETIC GRADUALISM; SPONTANEOUS ATAVISM

SALTATORY REPLICATION HYPOTHESIS [origin of repetitive DNA during evolution]

Britten, R. J., & D. E. Kohne. 1968. Repeated sequences in DNA. *Science* 161:529–40.

Tartof, K. D. 1975. Redundant genes. *Annu. Rev. Genet.* 9:355–85.

SALTING-OUT EFFECT [physiology]

Schmidt-Nielsen, K. 1979. *Animal Physiology: Adaptation and Environment.* Cambridge: Cambridge University Press, p. 431.

SARNIO'S LAWS [plant anatomy]

Sarnio, K. 1872. Über die Grösse der Holzzellen bei der gemeinen Kiefer *(Pinus silvestris). Jahrb. Wiss. Bot.* 8:401–20.

Bailey, I. W., & H. B. Shepard. 1915. Sarnio's laws for the variation in size of coniferous tracheids. *Bot. Gaz.* 60:66–71.

Carlquist, S. 1975. *Ecological Strategies of Xylem Evolution.* Berkeley: University of California Press, pp. 4*ff.*

SCALA NATURAE [evolution; philosophy]

Hodos, W., & C. B. G. Campbell. 1969. *Scala naturae:* why there is no theory in comparative psychology. *Psychol. Rev.* 76:337–50.

Martin, R. D. 1973. Comparative anatomy and primate systematics. *Symp. Zool. Soc. Lond.* 33:301–37.

Cf. CHAIN OF BEING

SCHOENER RULES [biogeography]

Willimas, E. E. 1972. The origin of faunas. Evolution of lizard congeners in a complex island fauna: a trial analysis. *Evol. Biol.* 6:47–89.

SCHÜRHOFF-BREWBAKER LAW [phylogenetic trends in the angiosperm male gametophyte]

Schürhoff, P. N. 1926. *Die Zytologie der Blütenpflanzen.* Stuttgart: Enke.

Brewbaker, J. L. 1967. The distribution and phylogenetic significance of binucleate and trinucleate pollen grains in the angiosperms. *Amer. J. Bot.* 54:1069–83.

Webster, G. L., & E. A. Rupert. 1973. Phylogenetic significance of pollen nuclear number in the Euphorbiaceae. *Evolution* 27:524–31.

SCINDULENE THEORY [biomembrane structure]

Nadol, J. B., J. R. Gibbins, & K. R. Porter. 1969. A reinterpretation of the structure and development of the basement lamella: an ordered array of collagen in fish skin. *Dev. Biol.* 20:304–31.

Løvtrup, S. 1974. *Epigenetics*. New York: Wiley, pp. 262–63.

SCRAMBLER HYPOTHESIS [of genetic recombination; immunology]

Smithies, O. 1967. Antibody variability. *Science* 157:267–73.

Burnet, F. M. 1969. *Cellular Immunology*. London: Cambridge University Press; Carlton: Melbourne University Press, p. 134.

SEARCH IMAGE = SEARCHING IMAGE *see* **SPECIFIC SEARCHING IMAGE**

SECOND SPECIES CONCEPT *see* **SPECIES CONCEPTS**

SECONDARY INTERGRADATION *see* **INTERGRADATION ZONES**

SEED PREDATION [ecology]

Janzen, D. H. 1970. Herbivores and the number of tree species in tropical forests. *Amer. Nat.* 104:501–28.

______ . 1971. Seed predation by animals. *Annu. Rev. Ecol. Syst.* 2:465–92.

Vandermeer, J. H. 1975. A graphical model of insect seed predation. *Amer. Nat.* 109:147–60.

SEGREGATION DISTORTION *see* **MEIOTIC DRIVE**

SEGREGATION, LAW OF *see* **MENDEL'S LAWS**

SELECTION HYPOTHESIS [evolution of genetic divergence]

Tartof, K. D. 1975. Redundant genes. *Annu. Rev. Genet.* 9:355–85.

SELECTION [evolution]

= selection theory

Weismann, A. 1909. The selection theory. *Darwin and Modern Science*. Edited by A. C. Seward. London: Cambridge University Press, pp. 18–65.

Lerner, I. M. 1958. *The Genetic Basis of Selection*. New York: Wiley.

Olson, E. C. 1960. Morphology, paleontology, and evolution.

Evolution After Darwin. Edited by S. Tax. Chicago: University of Chicago Press, vol. 1, pp. 523–45.

Cf. AGE-SPECIFIC SELECTION; CATASTROPHIC SELECTION; CLIMATIC SELECTION; DIRECTIONAL SELECTION; DISRUPTIVE SELECTION; ENDOCYCLIC SELECTION; GROUP SELECTION; NATURAL SELECTION; NORMALIZING SELECTION; PHYSIOLOGICAL SELECTION; SEXUAL SELECTION; STABILIZING SELECTION

SELECTIONIST VIEW [of biochemical evolution]

Clarke, B. 1970. Darwinian evolution of proteins. *Science* 168:1009–11.

Richmond, R. 1970. Non-Darwinian evolution: a critique. *Nature* 225:1025–28.

Milkman, R. 1972. How much room is left for non-Darwinian evolution? *Brookh. Symp. Biol.* 23:217–29.

Wills, C. 1973. In defense of naive pan-selectionism. *Amer. Nat.* 107:23–34.

Cf. NEUTRAL THEORY; NON-DARWINIAN EVOLUTION

SELECTIVE HYPOTHESES [of antibody formation; immunology]

Jerne, N. K. 1955. The natural-selection theory of antibody formation. *Proc. Natl. Acad. Sci.* 41:849–57.

Lederberg, J. 1959. Genes and antibodies. *Science* 129:1649–53.

Talmage, D. W. 1959. Immunological specificity. *Science* 129:1643–48.

Abramoff, P., & M. F. LaVia. 1970. *Biology of the Immune Response*. New York: McGraw-Hill, p. 304.

Cf. CLONAL SELECTION THEORY; EHRLICH'S SIDE CHAIN THEORY; IMMUNOLOGICAL TOLERANCE

SELECTIVE INERTIA [evolution]

Stebbins, G. L. 1968. Integration of development and evolutionary progress. *Population Biology and Evolution*. Edited by R. C. Lewontin. Syracuse, N.Y.: Syracuse University Press, pp. 17–36.

______ . 1974. *Flowering Plants: Evolution Above the Species Level*. Cambridge: Harvard University Press, p. 23.

SECONDARY THEOREM OF NATURAL SELECTION [evolution]

Robertson, A. 1968. The spectrum of genetic variation. *Population Biology and Evolution*. Edited by R. C. Lewontin.

Syracuse, N.Y.: Syracuse University Press, pp. 5–16.

Cf. FISHER'S FUNDAMENTAL THEOREM

SEMELPARITY *see* **BIG-BANG REPRODUCTION STRATEGY**

SEMISPECIES [systematics]

Mayr, E. 1931. Birds collected during the Whitney South Sea expedition. XII. Notes on *Halcyon chloris* and some of its subspecies. *Amer. Mus. Novit.* 469:1–10.

——— . 1963. *Animal Species and Evolution.* Cambridge: Belknap Press of Harvard University Press, p. 455.

Grant, V. 1963. *The Origin of Adaptations.* New York: Columbia University Press, p. 343.

———. 1971. *Plant Speciation.* New York: Columbia University Press, p. 47.

Baum, B. R. 1972. *Avena septentrionalis,* and the semispecies concept. *Can. J. Bot.* 50:2063–66.

Dobzhansky, T. 1972. Species of Drosophila: new excitement in an old field. *Science* 177:664–69.

———, F. J. Ayala, G. L. Stebbins, & J. W. Valentine. 1977. *Evolution.* San Francisco: Freeman, chaps. 6–7.

SENESCENCE [bioenergetics]

Weismann, A. 1891. *Essays on Heredity.* Oxford: Clarendon Press, chaps. 1, 3.

Bidder, G. P. 1932. Senescence. *Brit. Med. J.* 2:583–85.

Medawar, P. B. 1952. *An Unsolved Problem of Biology.* London: Lewis.

———. 1957. *The Uniqueness of the Individual.* London: Methuen.

Comfort, A. 1956. *The Biology of Senescence.* London: Routledge & Kegan Paul; New York: Rinehart.

Williams, G. C. 1957. Pleiotrophy, natural selection and the evolution of senescence. *Evolution* 11:398–411.

Leopold, A. C. 1975. Aging, senescence and turnover in plants. *Bioscience* 25:659–62.

Woolhouse, H. W. 1978. Senescence processes in the life cycle of flowering plants. *Bioscience* 28:25–31.

SENESCENCE [of species]

Beecher, C. E. 1898. The origin and significance of spines. A study in evolution. *Amer. J. Sci.* ser. 4, 6:1–20, 125–36, 249–68, 329–59.

Fernald, M. L. 1925. The persistence of plants in unglaciated areas of boreal America. *Mem. Amer. Acad. Arts Sci.* 15:239–342.

Cain, S. 1940. Some observations on the concept of species senescence. *Ecology* 21:213–15.

Romer, A. S. 1949. Time series and trends in animal evolution. *Genetics, Paleontology, and Evolution*. Edited by G. L. Jepson, E. Mayr, & G. G. Simpson. Princeton, N.J.: Princeton University Press, pp. 103–20.

SEQUENCE HYPOTHESIS [specificity of nucleic acids in DNA]

Crick, F. H. C. 1958. On protein synthesis. *Symp. Soc. Exp. Biol.* 12:138–63.

SEQUENCE THEORY [of parasite control]

Fiske, W. F. 1910. *Parasites of the Gypsy and Brown-Tail Moths Introduced into Massachusetts*. Boston: Wright & Potter.

Thompson, W. R. 1923. A criticism of the "sequence" theory of parasite control. *Ann. Entomol. Soc. Amer.* 16:115–28.

Cf. NATURAL CONTROL THEORIES

SERIAL ENDOSYMBIOSIS HYPOTHESIS [origin of eukaryotes]

Taylor, F. J. R. 1974. Implications and extensions of the serial endosymbiosis theory of the origin of eukaryotes. *Taxon* 23:229–58.

______ . 1976. Autogenous theories for the origin of eukaryotes. *Taxon* 25:377–90.

Cf. SYMBIOTIC THEORY

SEWALL WRIGHT EFFECT [genetics]

Fisher, R. A., & E. B. Ford. 1950. The "Sewall Wright effect." *Heredity* 4:117–19.

Wright, S. 1951. Fisher and Ford on "the Sewall Wright Effect." *Amer. Sci.* 39:452–58.

Cf. GENETIC DRIFT

SEXUAL SELECTION [evolution]

Darwin, C. 1859. *On the Origin of Species by Means of Natural Selection* . . . . London: Murray.

______. 1871. *The Descent of Man, and Selection in Relation to Sex*. London: Murray.

Fisher, R. A. 1930. *The Genetical Theory of Natural Selection*. Oxford: Clarendon Press, chap. 6.

Huxley, J. S. 1938. The present standing of the theory of sexual selection. *Evolution*. Edited by G. R. de Beer. Oxford: Clarendon Press, pp. 11–42.

O'Donald, P. 1962. The theory of sexual selection. *Heredity* 17:541–52.

Boesiger, E. 1967. La signification évolutive de la sélection sexuelle chez les animaux. *Scientia* 102:207–23.

Petit, C., & L. Ehrman. 1969. Sexual selection in *Drosophila*. *Evol. Biol.* 3:177–223.

Faugères, A., G. Petit, & E. Thibout. 1971. The components of sexual selection. *Evolution* 25:265–75.

Mayr, E. 1972. Sexual selection and natural selection. *Sexual Selection and the Descent of Man, 1871–1971.* Edited by B. G. Campbell. Chicago: Aldine, pp. 87–104.

O'Donald, P. 1977. Theoretical aspects of sexual selection. *Theor. Pop. Biol.* 12:298–334.

Willson, M. F. 1979. Sexual selection in plants. *Amer. Nat.* 113:777–90.

Cf. HANDICAP PRINCIPLE

SEXY SON HYPOTHESIS [of mate selection]

Weatherhead, P. J., & R. J. Robertson. 1979. Offspring quality and the polygyny threshold: the sexy son hypothesis. *Amer. Nat.* 113:201–08.

SHANNON-WEAVER DIVERSITY INDEX [ecology]

Shannon, C. E., & W. Weaver. *The Mathematical Theory of Communication*. Urbana: University of Illinois Press.

MacArthur, R. H. 1955. Fluctuations of animal populations and a measure of community stability. *Ecology* 36:533–36.

Pianka, E. R. 1966. Latitudinal gradients in species diversity: a review of concepts. *Amer. Nat.* 100:33–46.

Goodman, D. 1975. The theory of diversity-stability relationships in ecology. *Q. Rev. Biol.* 50:237–66.

Cf. DIVERSITY

SHELFORD'S LAW OF TOLERANCE *see* **TOLERANCE, LAW OF**

SHIFTING BALANCE THEORY [of evolution] *see* **BALANCE-SHIFT THEORY**

SHORT-SHOOT THEORY [of the ovuliferous scale; plant morphology]

Braun, A. 1853. Das Individuum der Pflanze. *Abh. Königl. Akad. Wiss. Berlin*, pp. 21–122.

Arber, A. 1950. *The Natural Philosophy of Plant Form*. Cambridge: Cambridge University Press, pp. 128*ff*.

SIBLING SPECIES [evolution]
= cryptic species
Mayr, E. 1942. *Systematics and the Origin of Species*. New York: Columbia University Press.
Burla, H., A. Brito da Cunha, A. R. Cordeiro, T. Dobzhansky, C. Malogolowkin, & C. Pavan. 1950. The *willistoni* group of sibling species of *Drosophila*. *Evolution* 3:300–14.
Dobzhansky, T. 1972. Species of Drosophila: new excitement in an old field. *Science* 177:664–69.

SIDE-CHAIN THEORY *see* **EHRLICH'S SIDE-CHAIN THEORY**

SIGN-STIMULI *see* **INNATE RELEASING MECHANISM (IRM)**

SIGNATURES, DOCTRINE OF [plants in medical superstition]
Porta, G. 1588. *Phytognomonica*. Naples: Horatium Salvianum.
Paracelsus. (1922–33.) *Samtliche Werke*. 14 vols. I. Abt. *Medizinische, naturwissenschaftliche und philosophische Schriften*. 14 vols. Edited by Karl Sudhoff. Munich: Oldenburg, vol. 13.
Arber, A. 1938. *Herbals, Their Origin and Evolution*. Cambridge: Cambridge University Press, p. 250.
Pagel, W. 1958. *Paracelsus*. Basel: Karger, p. 148.
Lehane, B. 1977. *The Power of Plants*. New York: McGraw-Hill, p. 152.

SIMILAR ECOLOGY CONCEPT
Gilbert, O., B. Reijnoldson, & J. Hobart. 1952. Gause's hypothesis: an examination. *J. Anim. Ecol.* 21:310–12.
Klomp, H. 1961. The concepts "similar ecology" and "competition" in animal ecology. *Arch. Néerl. Zool.* 14:90–102.

SIMILARITY AMONG DIVERSITY [taxonomy]
Sneath, P. H. A. 1971. Theoretical aspects of microbiological taxonomy. *Recent Advances in Microbiology*. Edited by A. Pérez-Miravete & D. Paláez. Mexico City: Asociacion Mexicana de Microbiologia, pp. 581–86.

SIMPLICITY ["Nature acts by simplest means"]
Blandino, G. 1969. *Theories on the Nature of Life*. New York: Philosophical Library, p. 260.
Cf. OCCAM'S RAZOR; PARSIMONY

SINGLE-ACTIVE X-CHROMOSOME HYPOTHESIS *see* **INACTIVE-X HYPOTHESIS**

SINGLE CELL THEORY [of the origin of eukaryotes]
Allsopp, A. 1969. Phylogenetic relationships of the Procaryota and the origin of the eukaryotic cell. *New Phytol.* 68:591–612.

SINGLE-GENE HETEROSIS *see* **CONDITIONAL HETEROSIS**

SINGLE LOCUS HETEROSIS *see* **CONDITIONAL HETEROSIS**

SITE-FAMILIARITY HYPOTHESIS [evolution of territoriality]
Carrick, R. 1963. Ecological significance of territory in the Australian magpie, *Gymnorhina tibicen*. *Proc. Int. Ornithol. Congr.* 13:740–53.
Verner, J. 1977. On the adaptive significance of territoriality. *Amer. Nat.* 111:769–75.

SIZE-EFFICIENCY HYPOTHESIS [in zooplankton; evolution]
Brooks, J. L., & S. I. Dodson. 1965. Predation, body size, and composition of plankton. *Science* 150:28–35.
Hall, D. J., & S. T. Threlkeld. 1976. The size-efficiency hypothesis and the size structure of zooplankton communities. *Annu. Rev. Ecol. Syst.* 7:177–208.

SLAVE-REPEATS HYPOTHESIS *see* **MASTER-SLAVE HYPOTHESIS**

SMALL ISLAND EFFECT [reduction in genetic variability]
Gorman, G. C., M. Soulé, & S. Y. Yang. 1975. Evolutionary genetics of insular adriatic lizards. *Evolution* 29:52–71.

SMITH-JONES HYPOTHESIS [primate relationships]
Smith, G. E. 1903. On the morphology of the brain in the Mammalia, with special reference to that of the lemurs, recent and extinct. *Trans. Linn. Soc. Lond.* 8:319–432.
Jones, F. W. 1916. *Arboreal Man*. London: Arnold.
Minkoff, E. C. 1974. The direction of lower primate evolution: an old hypothesis revived. *Amer. Nat.* 108:519–32.

SOCIAL DOMINANCE *see* **DOMINANCE** [behavior]

SOCIAL INERTIA [animal behavior]
Guhl, A. M. 1968. Social inertia and social stability in chickens. *Anim. Behav.* 16:219–32.

 SODIUM THEORY *see* **HODGKIN-HUXLEY AXON**

SOFT SELECTION *see* **HARD AND SOFT SELECTION**

SOIL-PLANT-ATMOSPHERE CONTINUUM [plant physiology]

Gradmann, H. 1928. Untersuchungen über die Wasserverhältnisse des Bodens als Grundlage des Pflanzenwachstums. *Jahrb.Wiss. Bot.* 69:1–100.

Van den Honert, T. H. 1948. Water transport as a catenary process. *Disc. Faraday Soc.* 3:288–98.

Philip, J. R. 1966. Plant water relations: some physical aspects. *Annu. Rev. Plant Physiol.* 17:245–68.

SOLAR-THERMAL UNIT THEORY or CONCEPT [ecology]

Caprio, J. M. 1971. The solar-thermal unit theory in relation to plant development and potential evapotranspiration. *Mont. Agri. Exp. Sta., Bozeman, Circ.* no. 251.

______ . 1974. The solar thermal unit concept in problems related to plant development and potential evapotranspiration. *Phenology and Seasonality Modeling*. Ecological Studies 8: *Synthesis and Modeling*. Edited by H. Lieth. New York: Springer-Verlag, pp. 353–64.

Cf. DEGREE DAY CONCEPT

SOMATIC CONVERSION *see* **PARAMUTATION**

SOMATIC MUTATIONAL HYPOTHESIS or THEORY [immunology]

= hypermutation theory

Brenner, S., & C. Milstein. 1966. Origin of antibody variation. *Nature* 211:242–43.

Abramoff, P., & M. F. LaVia. 1970. *Biology of the Immune Response*. New York: McGraw-Hill, p. 309.

Edelman, G. M. 1971. Antibody structure and molecular immunology. *Ann. N.Y. Acad. Sci.* 190:5–25.

Jerne, N. K. 1971. The somatic generation of immune recognition. *Eur. J. Immunol.* 1:1–9.

SPATIAL HETEROGENEITY THEORY [of species diversity]

Pianka, E. R. 1966. Latitudinal gradients in species diversity: a review of concepts. *Amer. Nat.* 100:33–46.

SPECIAL CREATION, DOCTRINE OF

= creationism

Anon. ? B.C. *Holy Bible*. part 1: Old Testament; book 1: Genesis. Various editions.

Brewster, E. T. 1927. *Creation: History of Non-Evolutionary Theories*. Indianapolis: Bobbs-Merrill.

Aulie, R. P. 1972. The doctrine of special creation. *Amer. Biol. Teach.* 34:191–200, 261–68, 281.

Newell, N. D. 1973. Special creation and organic evolution. *Proc. Amer. Phil. Soc.* 117:323–31.

Morris, H. M. 1974. *The Troubled Waters of Evolution*. San Diego: Creation-Life, pp. 104*ff*.

Hovenkamp, H. 1978. *Science and Religion in America, 1800–1860*. Philadelphia: University of Pennsylvania Press, pp. 119–45.

Cf. EVOLUTION

SPECIATION [evolution]

Mayr, E. 1963. *Animal Species and Evolution*. Cambridge: Belknap Press of Harvard University Press, chap. 15.

White, M. J. D. 1968. Models of speciation. *Science* 159:1065–70.

———. 1978. *Modes of Speciation*. San Francisco: Freeman.

Cf. ALLOPATRIC SPECIATION; APOMICTIC SPECIATION; CATASTROPHIC SPECIATION; CENTRIFUGAL SPECIATION; GEOGRAPHIC SPECIATION; HYBRID SPECIATION; ORTHOGENETIC SPECIATION; PARAPATRIC SPECIATION; QUANTUM SPECIATION; RECOMBINATIONAL SPECIATION; STASIPATRIC SPECIATION; SYMPATRIC SPECIATION

SPECIES CONCEPTS

Cain, S. A. 1944. *Foundations of Plant Geography*. New York: Harper, chap. 22.

Burma, B. H. 1949. The species concept: a semantic review. *Evolution* 3:369–70, 372–73.

Simpson, G. G. 1951. The species concept. *Evolution* 5:285–98.

Cain, A. J. 1953. Geography, ecology and coexistence in relation to the biological definition of the species. *Evolution* 7:76–83.

Meglitsch, P. A. 1954. On the nature of the species. *Syst. Zool.* 3:49–65.

Mayr, E., ed. 1957. *The Species Problem*. Washington, D. C.: American Association for the Advancement of Science, Publ. no. 50.

———. 1963. *Animal Species and Evolution*. Cambridge: Belknap Press of Harvard University Press, chap. 2.

———. 1968. Illiger and the biological species concept. *J. Hist. Biol.* 1:163–78.

———. 1969. The biological meaning of species. *Biol. J. Linn. Soc.* 1:311–20.

———. 1976. *Evolution and the Diversity of Life*. Cambridge: Belknap Press of Harvard University Press, pp. 481 *ff*.

Sokal, R. R., & P. H. A. Sneath. 1963. *Principles of Numerical Taxonomy*. San Francisco: Freeman.

Löve, A. 1965. The evolutionary framework of the biological species concept. *Genetics Today*. Edited by S. J. Geerts. Oxford: Pergamon, 2:409–15.

Lehman, H. 1967. Are biological species real? *Phil. Sci.* 34:157–67.

Larson, J. L. 1968. The species concept of Linnaeus. *Isis* 59:291–99.

Grant, V. 1971. *Plant Speciation*. New York: Columbia University Press, chaps. 2, 3*ff*.

Dobzhansky, T. 1972. Species of Drosophila: new excitement in an old field. *Science* 177:664–69.

Farber, P. L. 1972. Buffon and the concept of species. *J. Hist. Biol.* 5:259–84.

Grittenberger, E. 1972. Zum Artbegriff. *Acta Biotheor.* 21:47–62.

Sokal, R. R. 1973. The species problem reconsidered. *Syst. Zool.* 22:360–74.

Ghiselin, M. T. 1974. A radical solution to the species problem. *Syst. Zool.* 23:536–44.

Hull, D. L. 1976. Contemporary systematic philosophies. *Bost. Stud. Philos. Sci.* 27:396–440.

Mayr, E. 1976. Species concepts and definitions. *Bost. Stud. Philos. Sci.* 27:353–71.

Wiley, E. O. 1978. The evolutionary species concept reconsidered *Syst. Zool.* 27:17–26.

Løvtrup, S. 1979. The evolutionary species: fact or fiction? *Syst. Zool.* 28:386–92.

SPECIES-AREA CURVE [biogeography]

Jaccard, P. 1902. Lois de distribution florale dan la zone alpine. *Bull. Soc. Vaud. Sci. Nat.* 38:69–130.

Cain, S. A. 1938. The species-area curve. *Amer. Midl. Nat.* 19:573–81.

Goodall, D. W. 1952. Quantitative aspects of plant distribution. *Biol. Rev.* 27:194–245.

Preston, F. W. 1962. The canonical distribution of commonness and rarity. I. *Ecology* 43:185–215.

MacArthur, R. H., & E. O. Wilson. 1967. *The Theory of Island Biogeography*. Princeton, N.J.: Princeton University Press.

Haas, P. H. 1975. Some comments on the use of the species-area curve. *Amer. Nat.* 109:371–73.

Gould, S. J. 1979. An allometric interpretation of species-area curves: the meaning of the coefficient. *Amer. Nat.* 114:335-43.

Cf. AREA-PER SE HYPOTHESIS; ARRHENIUS'S EQUATION; CANONICAL DISTRIBUTION; ISLAND BIOGEOGRAPHY; RED QUEEN HYPOTHESIS

SPECIES-AREA THEORY [of insect-plant coevolution]

MacArthur, R. H., & E. O. Wilson. 1967. *The Theory of Island Biogeography*. Princeton, N.J.: Princeton University Press.

Claridge, M. F., & M. R. Wilson. 1978. British insects and trees: a study in island biogeography or insect/plant coevolution. *Amer. Nat.* 112:451–56.

SPECIES DIVERSITY *see* **DIVERSITY**

SPECIES PACKING [ecology]

Hutchinson, G. E. 1959. Homage to Santa Rosalia, or why are there so many kinds of animals? *Amer. Nat.* 93:145–59.

MacArthur, R. A. 1969. Species packing, or what competition minimizes. *Proc. Natl. Acad. Sci.* 64:1369–75.

______ . 1970. Species packing and competitive equilibrium for many species. *Theor. Pop. Biol.* 1:1–11.

MacArthur, R. H. 1972. *Geographical Ecology*. New York: Harper & Row, pp. 237–38.

May, R. M., & R. H. MacArthur. 1972. Niche overlap as a function of environmental variability. *Proc. Natl. Acad. Sci.* 69:1109–13.

Roughgarden, J., & M. Feldman. 1975. Species packing and predation pressure. *Ecology* 56:489–92.

Heck, K. L. 1976. Some critical considerations of the theory of species packing. *Evol. Theory* 1:247–58.

SPECIES PAIRS *see* **VICARIOUS SPECIES**

SPECIES PUMP HYPOTHESIS [biogeography]

Valentine, J. W. 1967. The influence of climatic fluctuations on species diversity within the tethyan provincial system. *Aspects of Tethyan Biogeography*. Edited by C. G. Adams & D. V. Ager. *Syst. Assoc. Publ.* 7:153–66.

Stebbins, G. L. 1974. *Flowering Plants: Evolution Above the Species Level*. Cambridge: Belknap Press of Harvard University Press, p. 194.

SPECIES SELECTION [evolution]

Stanley, S. M. 1975. A theory of evolution above the species

level. *Proc. Natl. Acad. Sci.* 72:646–50.
Cf. GEOGRAPHICAL SPECIATION; SALTATION

SPECIFIC DYNAMIC EFFECT [metabolic rate]
Rubner, M. 1902. *Die gesetze des energieverbrauchs bei der ernährung*. Leipzig & Wien: Deuticke.
Kleiber, M. 1961. *The Fire of Life*. New York: Wiley, p. 267.
Cf. RUBNER'S HYPOTHESIS; RUBNER'S LAW

SPECIFIC ENERGY OF THE NERVES, LAW OF *see* **MÜLLER'S LAW** [neuron activity]

SPECIFIC INDUCTION THEORY [of cellular interaction; morphogenesis]
Child, C. M. 1941. *Patterns and Problems of Development*. Chicago: University of Chicago Press.
Rose, S. M. 1957. Cellular interaction during differentiation. *Biol. Rev.* 32:351–82.

SPECIFIC INHIBITION THEORY [morphogenesis]
McCallum, W. B. 1905. Regeneration in plants. *Bot. Gaz.* 40:97–120, 241–63.
Rose, S. M. 1957. Cellular interaction during differentiation. *Biol. Rev.* 32:351–82.

SPECIFIC SEARCHING IMAGE [predation]
= search image
Tinbergen, L. 1960. The natural control of insects in pinewoods. I. Factors influencing the intensity of predation by songbirds. *Arch. Neerl. Zool.* 13:265–336.
Gibb, J. A. 1962. L. Tinbergen's hypothesis of specific search images. *Ibis* 104:106–11.
Croze, H. 1970. *Searching Image in Carrion Crows*. Berlin: Parey.
Royama, T. 1970. Factors governing the hunting behaviour and selection of food by the great tit (*Parus major* L.). *J. Anim. Ecol.* 39:619–68.
Mueller, H. 1974. Factors influencing prey selection in the American kestrel. *Auk* 91:705–21.

SPECIFIC SENSE ENERGIES *see* **MÜLLER'S LAW**

SPIEGELMAN'S RULE [development]
= limited realization, principle of
Spiegelman, S. 1945. Physiological competition as a regulatory mechanism in morphogenesis. *Q. Rev. Biol.* 20:121–46.

Wolpert, L. 1970. Positional information and pattern formation. *Towards a Theoretical Biology*. Edited by C. H. Waddington. Edinburgh: Edinburgh University Press, 3:198–230.

Cf. FRENCH FLAG

SPIRAL THEORY *see* **PHYLLOTAXIS**

SPONTANEOUS ATAVISM [evolution]

Riedl, R. 1977. A systems-analytical approach to macro-evolutionary phenomena. *Q. Rev. Biol.* 52:351–70.

Cf. RECAPITULATION; SALTATION

SPONTANEOUS GENERATION [origin of life]

Oparin, A. I. 1924. *Proiskhoyhdenie Zhizny*. Moscow: Izd. Moskovski Rabochii. (2d ed. transl. by S. Morgulis. 1938. *The Origin of Life*. London: Macmillan).

———. 1968. *Genesis and Evolutionary Development of Life*. New York: Academic Press.

Haldane, J. B. S. 1929. The origin of life. *Rationalist Annual*, 1929.

Conant, J. B., ed. 1967. *Pasteur's and Tyndall's Study of Spontaneous Generation*. Cambridge: Harvard University Press.

Vandervliet, G. 1971. *Microbiology and the Spontaneous Generation Debate During the 1870's*. Lawrence, Kans.: Coronado Press.

Farley, J. 1977. *The Spontaneous Generation Controversy from Descartes to Oparin*. Baltimore: Johns Hopkins Press.

Cf. ABIOGENESIS

SPREADING-OF-RISK THEORY [ecology; evolution]

den Boer, P. J. 1968. Spreading of risk and stabilization of animal numbers. *Acta Biotheor.* 18:165–94.

Reddingius, J., & P. J. den Boer. 1970. Simulation experiments illustrating stabilization of animal numbers by spreading of risk. *Oecologia* 5:240–84.

SQUARE LAW [of genetic equilibrium]

= CASTLE'S LAW = HARDY-WEINBERG LAW

Spiess, E. B. 1977. *Genes in Populations*. New York: Wiley, pp. 22*ff*.

STABILITY [ecological]

Diversity and Stability in Ecological Systems. (1969.) *Brookh. Symp. Biol.*, no. 22.

Holling, C. S. 1973. Resilience and stability of ecological systems. *Annu. Rev. Ecol. Syst.* 4:1-23.

May, R. M. 1973. *Stability and Complexity in Model Ecosystems.* Princeton, N.J.: Princeton University Press.

______ . 1975. Stability in ecosystems: some comments. *Unifying Concepts in Ecology.* Edited by W. H. van Dobben & R. H. Lowe-McConnell. The Hague: Junk, pp. 161–68.

Usher, M. B., & M. H. Williamson. 1974. *Ecological Stability.* New York: Halsted.

Orians, G. H. 1975. Diversity, stability and maturity in natural ecosystems. *Unifying Concepts in Ecology.* Edited by W. H. van Dobben & R. H. Lowe-McConnell. The Hague: Junk, pp. 139–50.

Cf. DIVERSITY

STABILITY-TIME HYPOTHESIS [patterns of species diversity]

Sanders, H. L. 1968. Marine benthic diversity: a comparative study. *Amer. Nat.* 102:243–82.

______ . 1969. Benthic marine diversity and the stability-time hypothesis. *Brookh. Symp. Biol.* 22:71–80.

Slobodkin, L. B., & H. L. Sanders. 1969. On the contribution of environmental predictability to species diversity. *Brookh. Symp. Biol.* 22:82–95.

Ayala, F. J., J. W. Valentine, D. Hedgecock, & L. G. Barr. 1975. Deep-sea asteroids: high genetic variability in a stable environment. *Evolution* 29:203–12.

Menge, B. A., & J. P. Sutherland. 1976. Species diversity gradients: synthesis of the roles of predation, competition, and temporal heterogeneity. *Amer. Nat.* 110:351–69.

Abele, L. G., & K. Walters. 1979. The stability-time hypothesis: reevaluation of the data. *Amer. Nat.* 114:559–68.

STABILIZING SELECTION

= NORMALIZING SELECTION = centripetal selection

Schmalhausen, I. I. 1949. *Factors of Evolution. The Theory of Stabilizing Selection.* Philadelphia: Blakiston.

Mather, K. 1953. The genetical structure of populations. *Symp. Soc. Exp. Biol.* 7:66–95.

Simpson, G. G. 1953. *The Major Features of Evolution.* New York: Columbia University Press, p. 148.

Cf. DISRUPTIVE SELECTION

STABLE POPULATION THEORY [ecology]

Lotka, A. J. 1924. *Elements of Physical Biology.* Baltimore: Williams & Wilkins.

Leslie, P. H. 1945. On the use of matrices in certain population mathematics. *Biometrika* 35:183–212.

Gourley, G. S., & C. E. Lawrence. 1977. Stable population analysis in periodic environments. *Theor. Pop. Biol.* 11:49–59.

STARLING'S LAW OF THE HEART [physiology]

The Physiological Basis of Starling's Law of the Heart. 1974. (Ciba Foundation Symposium 24.) Amsterdam: Associated Scientific Publishers.

Schmidt-Nielsen, K. 1979. *Animal Physiology: Adaptation and Environment.* Cambridge: Cambridge University Press, p. 103.

STASIGENESIS [evolution]

Huxley, J. S. 1957. The three types of evolutionary process. *Nature* 180:454–55.

STASIPATRIC SPECIATION [evolution]
= PARAPATRIC SPECIATION

White, M. J. D., R. E. Blackith, R. M. Blackith, & J. Cheney. 1967. Cytogenetics of the *viatica* group of Morabine grasshoppers. I. The "coastal" species. *Aust. J. Zool.* 15:263–302.

White, M. J. D. 1968. Models of speciation. *Science* 159:1065–70.

______ . 1973. *Animal Cytology and Evolution.* 3d ed. Cambridge: Cambridge University Press, chap. 11.

STELAR THEORY [plant anatomy]

Van Tieghem, P. 1886. Sur la polystélie. *Ann. Sci. Nat. Bot.* Ser. 7, 3:275–322.

Bower, F. O. 1923. *The Ferns (Filicales).* Vol. I: *Analytical Examination of the Criteria of Comparison.* Cambridge: Cambridge University Press, p. 138.

Carlquist, S. 1975. *Ecological Strategies of Xylem Evolution.* Berkeley: University of California Press, p. 61.

STENSIÖ-ØRVIG LEPIDOMORIAL THEORY *see* **LEPIDOMORIAL THEORY**

STEPPING-STONE MODEL [of genetic variation]

Kimura, M. 1953. "Stepping stone" model of population. *Annu. Rep. Natl. Inst. Genet.* 3:62–63.

Maruyama, T. 1970. On the rate of decrease of heterozygosity in circular stepping stone models of population. *Theor. Pop. Biol.* 1:101–19.

______. 1971. Speed of gene substitution in a geographically structured population. *Amer. Nat.* 105:253–65.

Cf. NEUTRAL THEORY

STEPWISE DETERMINATION [of vertebrate development]

Hadorn, E. 1965. Problems of determination and trans-determination. *Brookh. Symp. Biol.* 18:148–61.

Gehring, W. J. 1976. Determination of primordial disc cells and the hypothesis of stepwise determination. *Symp. Roy. Entomol. Soc. Lond.* 8:99–108.

STEREOCHEMICAL THEORY [origin and universality of the genetic code]

Crick, F. H. C. 1968. The origin of the genetic code. *J. Mol. Biol.* 38:367–79.

STOCHASTIC THEORY [of gene frequency]

= stochastic variability

Fisher, R. A. 1930. *The Genetical Theory of Natural Selection.* Oxford: Clarendon Press.

Wright, S. 1931. Evolution in Mendelian populations. *Genetics* 16:97–159.

Kimura, M. 1955. Stochastic processes and distribution of gene frequencies under natural selection. *Cold Spr. Harb. Symp. Quant. Biol.* 20:33–53.

Milkman, R. 1970. The genetic basis of natural variation in *Drosophila melanogaster. Adv. Genet.* 15:55–114.

Gillespie, J. H. 1972. The effects of stochastic environments on allele frequencies in natural populations. *Theor. Pop. Biol.* 3:241–48.

STOCKARD'S FOUR PRINCIPLES [of teratogenesis]

Stockard, C. R. 1921. Developmental rate and structural expression. *Amer. J. Anat.* 28:115–277.

Hughes, A. F. W. 1976. Developmental biology and the study of malformations. *Biol. Rev.* 51:143–79.

STOKE'S LAW [pollen dispersal]

Schmidt, W. 1918. Die Verbreitung von Samen und Blutenstaub durch die Luftbewegung. *Österr. Bot. Z.* 67:313–28.

Levin, D. A., & H. W. Kerster. 1974. Gene flow in seed plants. *Evol. Biol.* 7:139–220.

STORAGE THEORY [of evolutionary change]

Mayr, E. 1963. *Animal Species and Evolution.* Cambridge:

Belknap Press of Harvard University Press, chap. 9.
Riedl, R. 1977. A systems-analytical approach to macro-evolutionary phenomena. *Q. Rev. Biol.* 52:351–70.

STRATEGY [ecology]
Harper, J. L., & J. Ogden. 1970. The reproductive strategy of higher plants. I. The concept of strategy with special reference to *Senecio vulgaris* L. *J. Ecol.* 58:681–98.

STRATEGY [evolution]
Slobodkin, L. B. 1964. The strategy of evolution. *Amer. Sci.* 52:342–57.

STRESS SYNDROME [physiology; ecology]
Selye, H. 1956. *The Stress of Life*. New York: McGraw-Hill.
Klopfer, P. H. 1973. *Behavioral Aspects of Ecology*. 2d ed. Englewood Cliffs, N.J.: Prentice-Hall, p.59.

STRESS-TOLERANT STRATEGY [in plants]
Grime, J. P. 1977. Evidence for the existence of three primary strategies in plants and its relevance to ecological and evolutionary theory. *Amer. Nat.* 111:1169–94.

STROBILAR THEORY [of shoot organization in plants] *see* **AXIAL THEORIES**

STRUGGLE FOR EXISTENCE [Darwinian evolution]
Darwin, C. 1859. *On the Origin of Species by Means of Natural Selection* . . . London: Murray, p. 62.
Gale, B. G. 1972. Darwin and the concept of a struggle for existence: a study in the extrascientific origins of scientific ideas. *Isis* 63:321–44.

SUBNUCLEAR HYPOTHESIS [of macronuclear organization; morphogenesis; genetics]
Sonneborn, T. M. 1947. Recent advances in the genetics of *Paramecium* and *Euplotes*. *Adv. Genet.* 1:263–358.
Nanney, D. L. 1964. Macronuclear differentiation and subnuclear assortment in ciliates. *Role of Chromosomes in Development*. Edited by M. Locke. New York: Academic Press, pp. 253–73.
Raikov, I. B. 1976. Evolution of macronuclear organization. *Ann. Rev. Genet.* 10:413–40.
Cf. COMPOSITE CHROMOSOME THEORY; MASTER-SLAVE HYPOTHESIS; NUCLEOSOME HYPOTHESIS

SUBSIDENCE THEORY [of coral reef formation]

Darwin, C. 1842. *The Structure and Distribution of Coral Reefs*. London: Smith, Elder.

Stoddart, D. R. 1969. Ecology and morphology of recent coral reefs. *Biol. Rev.* 44:433–98.

SUBSTITUTIONAL LOAD [population fitness]

Kimura, M. 1960. Genetic load of a population and its significance in evolution. *Jap. J. Genet.* 35:7–33.

______ . 1960. Optimum mutation rate and degree of dominance as determined by the principle of minimum genetic load. *J. Genet.* 57:21–34.

______ . 1961. Natural selection as the process of accumulating genetic information in adaptive evolution. *Genet. Res.* 2:127–40.

Cf. HALDANE'S DILEMMA

SUCCESSION [ecology]

Cowles, H. C. 1901. The physiographic ecology of Chicago and vicinity. *Bot. Gaz.* 31:73–108, 145–82.

Gleason, H. A. 1910. The vegetation of the inland sand deposits of Illinois. *Bull. Ill. St. Lab. Nat. Hist.* 9:21–174.

Clements, F. E. 1916. Plant succession. *Carnegie Inst. Wash. Publ.* 242.

Gleason, H. A. 1927. Further views on the succession concept. *Ecology* 8:299–326.

Tansley, A. G. 1929. Succession, the concept and its values. *Proceedings of the International Congress of Plant Sciences*, Ithaca, N.Y., Aug. 16–23, 1926. Edited by B. M. Duggar. 2 vols. Menasha, Wis.: Banta. 1:677–86.

Drury, W. H., & I. C. T. Nisbet. 1973. Succession. *J. Arnold Arbor.* 54:331–68.

McIntosh, R. P. 1975. H. A. Gleason—"individualistic ecologist" 1882–1975: his contributions to ecological theory. *Bull. Torr. Bot. Club* 102:253–73.

Pickett, S. T. A. 1976. Succession: an evolutionary interpretation. *Amer. Nat.* 110:107–19.

Connell, J. H., & R. O. Slayter. 1977. Mechanisms of succession in natural communities and their role in community stability and organization. *Amer. Nat.* 111:1119–44.

Cf. ADJUSTMENT STABILITY; CYCLIC SUCCESSION

SUFFICIENT-RESOURCE HYPOTHESIS [evolution of territoriality]

Verner, J. 1977. On the adaptive significance of territoriality. *Amer. Nat.* 111:769–75.

SUPER-GENE [genetics]

Darlington, C. D., & K. Mather. 1949. *The Elements of Genetics*. London: Allen & Unwin, pp. 336*ff*.

Sheppard, P. M. 1953. Polymorphism, linkage and the blood groups. *Amer. Nat.* 87:238–94.

______. 1969. Evolutionary genetics of animal populations: the study of natural populations. *Proc. XII Int. Congr. Genet.* 3:261–79.

Mather, K. 1955. Polymorphism as an outcome of disruptive selection. *Evolution* 9:52–61.

Ford, E. B. 1965. *Genetic Polymorphism*. Cambridge: M.I.T. Press.

______. 1971. *Ecological Genetics*. New York: Wiley.

______. 1974. Supergenes: are these ecological operons? *Brookh. Symp. Biol.* 25:297–308.

Clarke, C. A., & P. M. Sheppard. 1960. Super-genes and mimicry. *Heredity* 14:175–85.

Turner, J. R. G. 1967. On supergenes. I. The evolution of supergenes. *Amer. Nat.* 101:195–221.

______. 1977. Butterfly mimicry: the genetical evolution of an adaptation. *Evol. Biol.* 10:163–206.

SUPERNORMAL STIMULI [behavior]

Koehler, O. 1932. Beiträge zur Sinnephysiologie der Süsswasserplanarian. *Z. Vergl. Physiol.* 16:606–756.

Tinbergen, N. 1948. Social releasers and the experimental method required for their study. *Wilson Bull.* 60:6–51.

______. 1970. *The Study of Instinct*. Oxford: Oxford University Press, pp. 44–46.

Staddon, J. E. R. 1975. A note on the evolutionary significance of "supernormal" stimuli. *Amer. Nat.* 109:541–45.

Cf. INNATE RELEASING MECHANISMS

SUPERSPECIES [systematics; evolution]

Rensch, B. 1929. *Das Prinzip geographischer Rassenkreise und das Problem der Artbildung*. Berlin: Gebrüder Borntraeger.

Mayr, E. 1963. *Animal Species and Evolution*. Cambridge: Belknap Press of Harvard University Press, p. 672.

______. 1969. *Principles of Systematic Zoology*. New York: McGraw-Hill, p. 52.

Amadon, D. 1966. The superspecies concept. *Syst. Zool.* 15:245–49.

Dobzhansky, T. 1972. Species of Drosophila: new excitement

in an old field. *Science* 177:664–69.

Winterbottom, J. M. 1972. Comments on the superspecies concept. *Ibis* 114:401–03.

Van Bemmel, A. C. V. 1973. The concept of superspecies applied to Eurasiatic Cervidae. *Z. Säugetierk.* 38:295–302.

SUPER-TERRITORY HYPOTHESIS [evolution of territoriality]

Verner, J. 1977. On the adaptive significance of territoriality. *Amer. Nat.* 111:769–75.

Colgan, P. 1979. Is a super-territory strategy stable? *Amer. Nat.* 114:604–05.

Getty, T. 1979. On the benefits of aggression: the adaptiveness of inhibition and super territories. *Amer. Nat.* 114:605–09.

Pleasants, J. M., & B. Y. Pleasants. 1979. The super-territory hypothesis: a critique, or why there are so few bullies. *Amer. Nat.* 114:609–14.

SUPRAORGANISMIC CONCEPT [of the plant community]

= Clementsian organismic view

Clements, F. E. 1905. *Research Methods in Ecology*. Lincoln, Neb.: University Publishing.

______ . 1916. Plant succession: an analysis of the development of vegetation. *Carnegie Inst. Wash. Publ.* 242.

Gleason, H. A. 1917. The structure and development of the plant association. *Bull. Torr. Bot. Club.* 44:463–81.

McIntosh, R. P. 1975. H. A. Gleason—"individualistic ecologist" 1882–1975: his contributions to ecological theory. *Bull. Torr. Bot. Club* 102:253–73.

Cf. INDIVIDUALISTIC CONCEPT

SURFACE LAW [metabolic rate: surface area]

= surface rule = Rubner's rule

Sarrus, & J. F. Rameaux. 1839. Mémoire adressé à l'Académie Royale. *Bull. Acad. Roy. Med.* 3:1094–100.

Rubner, M. 1883. Ueber den Einfluss der Körpergrösse auf Stoff- und Kraftwechsel. *Z. Biol.* 19:535–62.

Kleiber, M. 1947. Body size and metabolic rate. *Physiol. Rev.* 27:511–41.

______ . 1961. *The Fire of Life*. New York: Wiley, chaps. 10, 13.

Bertalanffy, L. von. 1951. Metabolic types and growth rates. *Amer. Nat.* 85:111–17.

Cf. RUBNER'S LAW; SPECIFIC DYNAMIC EFFECT

SURFACE TENSION THEORY [of protoplasmic movement]

Thompson, D'A. W. 1942. *On Growth and Form*. 2d ed. Cambridge: Cambridge University Press, pp. 360–61.

SUSPENDED ANIMATION *see* **ANABIOSIS**

SUTTON-BOVERI THEORY OF CHROMOSOMAL INHERITANCE *see* **CHROMOSOMAL THEORY OF HEREDITY**

SUTURE ZONES [biogeography]
= HYBRID ZONES = INTERGRADATION ZONES

Remington, C. L. 1968. Suture-zones of hybrid interaction between recently joined biotas. *Evol. Biol.* 2:321–428.

Short, L. L. 1969. "Suture zones," secondary contacts, and hybridization. *Syst. Zool.* 18:458–60.

Uzzell, T., & N. P. Ashmole. 1970. Suture zones: an alternative view. *Syst. Zool.* 19:197–99.

SYMBIOSIS [physiology; evolution]

de Bary, A. 1863. Recherches sur le développement de quelques champignons parasites. *Ann. Sci. Nat. Bot.*, ser. 4, 20:5–148.

______. 1879. *Die Erscheinung der Symbiose.* Strassburg: Trübner.

______. 1887. *Comparative Morphology and Biology of the Fungi, Mycetozoa and Bacteria.* Oxford: Clarendon Press, p. 356.

Hertig, M., W. H. Taliafirro, & B. Schwartz. 1937. The terms symbiosis, symbiont and symbiote. *J. Parasit.* 23:326–29.

Lewis, D. H. 1973. Concepts in fungal nutrition and the origin of biotrophy. *Biol. Rev.* 48:261–78.

Cf. BIOTROPHY; MUTUALISM

SYMBIOTIC THEORY [of the origin of eukaryotic cells]

Mereschkowsky, C. 1905. Über Natur und Orsprung der Chromatophoren im Pflanzenreiche. *Biol. Zblt.* 25:593–604.

Echlin, P. 1966. The blue-green algae. *Sci. Amer.* 214:74–81.

Sagan, L. 1967. On the origin of mitosing cells. *J. Theor. Biol.* 14:225–74.

Stanier, R. Y. 1970. Some aspects of the biology of cells and their possible evolutionary significance. *Symp. Soc. Gen. Microbiol.* 20:1–38.

Margulis, L. 1970. *Origin of Eukaryotic Cells.* New Haven, Conn.: Yale University Press.

______. 1974. Five-kingdom classification and the origin and evolution of cells. *Evol. Biol.* 7:45–78.

______. 1975. Symbiotic theory of the origin of eukaryotic

organelles: criteria for proof. *Symp. Soc. Exp. Biol.* 29:21–38.

Raff, R. A., & H. R. Mahler. 1972. The non symbiotic origin of mitochondria. *Science* 177:575–82.

Ragan, M. A., & D. J. Chapman. 1978. *A Biochemical Phylogeny of the Protists.* New York: Academic Press.

Cf. CLUSTER CLONE THEORY; INVAGINATION THEORY; NON-SYMBIOTIC THEORY

SYMPATRIC SPECIATION [evolution]

Mayr, E. 1942. *Systematics and the Origin of Species.* New York: Columbia University Press, pp. 189*ff.*

Thoday, J. M., & T. B. Boam. 1959. Effects of disruptive selection. II. Polymorphism and divergence without isolation. *Heredity* 13:205–18.

———, & J. B. Gibson. 1962. Isolation by disruptive selection. *Nature* 193:1164–66.

———. 1970. The probability of isolation by disruptive selection. *Amer. Nat.* 104:219–30.

Maynard Smith, J. 1962. Disruptive selection, polymorphism, and sympatric speciation. *Nature* 195:60–62.

———. 1966. Sympatric speciation. *Amer. Nat.* 100:637–50.

Pimentel, D., G. J. C. Smith, & J. Soans. 1967. A population model of sympatric speciation. *Amer. Nat.* 101:493–504.

SYNORGANIZATION [evolution]

= COADAPTATION

Remane, A. 1971. *Die Grundlagen des natürlichen Systems der vergleichenden Anatomie und der Phylogenetik.* Könnigstein-Taunus: Koeltz.

Osche, G. 1972. *Evolution, Grundlagen—Erkentrisse—Entwicklungen der Abstammungslehre.* Basel: Herder.

Riedl, R. 1977. A systems-analytical approach to macro-evolutionary phenomena. *Q. Rev. Biol.* 52:351–70.

SYNTENY [genetics]

Renwick, H. H. 1971. The mapping of human chromosomes. *Ann. Rev. Genet.* 5:81–120.

Novitski, E., & S. Blixt. 1978. Mendel, linkage, and synteny. *Bioscience* 28:34–35.

SYNTHETIC THEORY OF CLASSIFICATION *see* **EVOLUTIONARY SYSTEMATICS**

SYNTHETIC THEORY [of evolution]

= NEO-DARWINIAN EVOLUTION = biological theory of evolution = modern synthesis

Tshetverikov, S. S. 1926. On certain aspects of the evolutionary process from the standpoint of modern genetics. Transl. by M. Barker. *Proc. Amer. Philos. Soc.* 105:167–95.

Dobzhansky, T. 1967. Sergei Sergeerich Tshetverikov (1880–1959). *Genetics* 55:1–3.

______, F. J. Ayala, G. L. Stebbins, & J. W. Valentine. 1977. *Evolution.* San Francisco: Freeman, pp. 5*ff*.

Huxley, J. S. 1942. *Evolution, The Modern Synthesis.* New York: Harper.

SYSTEM THEORY [of evolution]

Riedl, R. 1977. A systems-analytical approach to macro-evolutionary phenomena. *Q. Rev. Biol.* 52:351–70.

SYSTEMATIC FITNESS *see* **INCLUSIVE FITNESS**

SZIDAT'S RULE [more specialized hosts have more specialized parasites]

Eichler, W. 1948. Some rules in ectoparasitism. *Ann. Mag. Nat. Hist.*, ser. 12, 1:588–98.

Janiszewska, J. 1949. Parasitogenetic rules. Janicki rule. *Zool. Pol.* 5:31–34.

Szidat, L. 1956. Geschichte, anwendung und einige folgerungen aus den parasitogenetischen regeln. *Z. Parasitenk.* 17:237–68.

Price, P. W. 1977. General concepts on the evolutionary biology of parasites. *Evolution* 31:405–20.

[T]

TACHYTELY [evolutionary rate distribution]

= a-rates

Simpson, G. G. 1944. *Tempo and Mode in Evolution.* New York: Columbia University Press, chap. 4.

______. 1953. *The Major Features of Evolution.* New York: Columbia University Press, chap. 10.

Kurtén, B. 1959. Rates of evolution in fossil mammals. *Cold Spr. Harb. Symp. Quant. Biol.* 24:205–15.

Cf. BRADYTELY; HOROTELY

TARGET THEORY [radiation absorption to estimate gene size]

Crowther, J. A. 1924. Some considerations relative to the action of x-rays on tissue cells. *Proc. Roy. Soc. Lond. B* 96:207–11.

———. 1927. A theory of the action of x-rays on living cells. *Proc. Camb. Philos. Soc.* 23:284–87.

Hollaender, A. 1961. Hit and target theories. *Science* 134:1233.

Carlson, E. A. 1966. *The Gene: A Critical History.* Philadelphia: Saunders, pp. 158–65.

Zimmer, K. G. 1966. The target theory. *Phage and the Origins of Molecular Biology.* Edited by J. Cairns, G. Stent, & J. D. Watson. Cold Spring Harbor, N.Y.: Cold Spring Harbor Laboratory of Quantitative Biology, pp. 33–42.

Cf. HIT THEORY

TAXON CYCLE [patterns of expansion and restriction in taxa]

Wilson, E. O. 1961. The nature of the taxon cycle in the Melanesian ant fauna. *Amer. Nat.* 95:169–93.

Ricklefs, R. E., & G. W. Cox. 1972. Taxon cycles in the West Indian avifauna. *Amer. Nat.* 106:195–219.

Pianka, E. R. 1978. *Evolutionary Ecology.* 2d ed. New York: Harper & Row, p. 326.

TAXONOMIC SPACE [numerical taxonomy]

Silvestri, L., M. Turri, L. R. Hill, & E. Gilardi. 1962. A quantitative approach to the systematics of Actinomycetes, based on overall similarity. *Symp. Soc. Gen. Microbiol.* 12:333–60.

Hutchinson, G. E. 1968. When are species necessary? *Population Biology and Evolution.* Edited by R. C. Lewontin. Syracuse, N.Y.: Syracuse University Press, pp. 177–86.

TELOME THEORY [morphology; evolution]

Zimmerman, W. 1930. *Phylogenie der Pflanzen: Ein Überblick über Tatsachen und Probleme.* Jena: Fischer.

———. 1959. *Die Phylogenie der Pflanzen.* 2d ed. Stuttgart: Fischer.

———. 1965. Die Blütenstände, ihr System und ihre Phylogenie. *Ber. Deut. Bot. Ges.* 78:3–12.

Wardlaw, C. W. 1952. *Phylogeny and Morphogenesis.* London: Macmillan, pp. 109–12.

TELEONOMY [study of adaptation]

Pittendrigh, C. S. 1958. Adaptation, natural selection, and behavior. *Behavior and Evolution.* Edited by A. Roe & G. G. Simpson. New Haven, Conn.: Yale University Press, pp. 390–416.

Williams, G. C. 1966. *Adaptation and Natural Selection: A Critique of Some Current Evolutionary Thought.* Princeton, N.J.: Princeton University Press, chap. 9.

Cf. ADAPTATION

TEMPLATE CONCEPT [of gene replication]

Pontecorvo, G. 1966. Template and stepwise processes in heredity. *Proc. Roy. Soc. Lond. B* 164:167–69.

Olby, R. 1971. Schrödinger's problem: what is life? *J. Hist. Biol.* 4:119–48.

TEMPLATE HYPOTHESES or THEORIES [of antibody formation; immunology]

Breinl, F., & F. Haurowitz. 1930. Chemische Untersuchung des Praecipitates aus Haemoglobin und Anti-Haemoglobin-serum and Bemerkungen ueber die Natur der Antikoerper. *Hoppe-Seyl. Z.* 192:45–57.

Alexander, J. 1932. Some intracellular aspects of life and disease. *Protoplasma* 14:296–306.

Mudd, S. 1932. A hypothetical mechanism of antibody formation. *J. Immunol.* 23:423–27.

Pauling, L. 1940. A theory of the structure and process of formation of antibodies. *J. Amer. Chem. Soc.* 62:2643–57.

Haurowitz, F. 1952. The mechanism of the immunological response. *Biol. Rev.* 27:247–80.

Abramoff, P., & M. F. LaVia. 1970. *Biology of the Immune Response*. New York: McGraw-Hill, p. 304.

TEMPLATE THEORY [protein synthesis]

Haurowitz, F. 1949. Biological problems and immunochemistry. *Q. Rev. Biol.* 24:93–123.

Olby, R. 1974. The origins of molecular genetics. *J. Hist. Biol.* 7:93–100.

Portugal, F. H., & J. S. Cohen. 1977. *A Century of DNA*. Cambridge: M.I.T. Press, chap. 12.

TENSION ZONE [ecology]

Griggs, R. F. 1914. Observations on the behavior of some species at the edges of their ranges. *Bull. Torr. Bot. Club* 41:25–49.

Weaver, J. E., & A. F. Thiel. 1917. Ecological studies in the tension zone between prairie and woodland. *Botanical Survey of Nebraska* (n.s.) no. 1.

Pool, R. J., J. E. Weaver, & F. C. Jean. 1918. Further studies in the ecotone between prairie and woodland. *University of Nebraska Studies* 18:7–47.

Curtis, J. T. 1959. *The Vegetation of Wisconsin*. Madison: University of Wisconsin Press, pp. 15*ff*.

TERATOLOGY, DOCTRINE OF [plant evolution]

Worsdell, W. C. 1915–1916. *The Principles of Plant*

Teratology. 2 vols. London: Ray Society.

Heslop-Harrison, J. 1952. A reconsideration of plant teratology. *Phyton* 4:19–34.

Sporne, K. R. 1956. The phylogenetic classification of the angiosperms. *Biol. Rev.* 31:1–29.

Cf. RECAPITULATION

TERRITORIALITY [behavior]

= territory

Howard, E. 1923. *Territory in Bird Life*. London: Murray.

Nice, M. M. 1933. The theory of territorialism and its development. *Fifty Years' Progress of American Ornithology, 1883-1933*. Edited by F. M. Chapman & T. S. Palmer. Lancaster, Pa.: American Ornithological Union, pp. 89–100.

______. 1941. The role of territoriality in bird life. *Amer. Midl. Nat.* 26:441–87.

______. 1953. The earliest mention of territory. *Condor* 55:316–17.

Stokes, A. W., ed. *Territory*. Stroudsburg, Pa.: Dowden, Hutchinson & Ross.

Wilson, E. O. 1975. *Sociobiology*. Cambridge: Belknap Press of Harvard University Press, pp. 260*ff*.

Hutchinson, G. E. 1978. *An Introduction to Population Ecology*. New Haven, Conn.: Yale University Press, pp. 105 *ff*.

Cf. HOME RANGE

TETRANUCLEOTIDE HYPOTHESIS [structure of nucleic acids]

Kossel, A., & A. Neumann. 1893. Ueber das Thymin, ein Spaltungsprodukt der Nucleinsäure. *Ber. Deut. Chem. Ges.* 26:2753–56.

Levene, P. A. 1921. On the structure of thymus nucleic acid and on its possible bearing on the structure of plant nucleic acid. *J. Biol. Chem.* 48:119–25.

Olby, R. 1974. DNA before Watson–Crick. *Nature* 248:782–85.

Portugal, F. H., & J. S. Cohen. 1977. *A Century of DNA*. Cambridge: M.I.T. Press, chap. 4

THAYER'S LAW [of animal coloration]

Thayer, A. H. 1896. The law which underlies protective coloration. *Auk* 13:124–219, 318–20.

Kingsland, S. 1978. Abbott Thayer and the protective coloration debate. *J. Hist. Biol.* 11:223–44.

THERMAL THEORY [of biochemical origins]

Fox, S. W. 1960. How did life begin? *Science* 132:200–08.

———. 1963. Experiments suggesting origins of amino acids and proteins. *Protein Nutrition and Metabolism*. Edited by J. Kastelic, H. H. Draper, & H. P. Broquist. Urbana: Spec. Publ. no. 4 of the University of Illinois College of Agriculture, pp. 141–54.

THIRD LAW OF NATURAL SELECTION

Van Valen, L. 1973. A new evolutionary law. *Evol. Theory* 1:1–30.

———. 1976. Energy and evolution. *Evol. Theory* 1:179–229.

THIRD SPECIES CONCEPT *see* **SPECIES CONCEPTS**

THOM'S THEORY OF CATASTROPHES [morphogenesis; evolution]

Thom, R. 1969. Topological models in biology. *Topology* 8:313–35.

———. 1970. Topological models in biology. *Towards a Theoretical Biology*. Edited by C. H. Waddington, Edinburgh: Edinburgh University Press, 3:89–116.

———. 1974. *Structural Stability and Morphogenesis*. Reading, Mass.: Benjamin.

Stewart, I. N. 1975. The seven elementary catastrophes. *New Sci.* 68:447–54.

Dodson, M. M. 1976. Darwin's law of natural selection and Thom's theory of catastrophes. *Math. Biosci.* 28:243–74.

THOMPSON'S THEORY *see* **NATURAL CONTROL THEORIES**

THRESHOLD SELECTION (1) [evolution]

Maynard Smith, J. 1968. "Haldane's dilemma" and the rate of evolution. *Science* 219:1114–16.

Nei, M. 1971. Fertility excess necessary for gene substitution in regulated populations. *Genetics* 68:169–84.

THRESHOLD SELECTION (2) *see* **GENETIC ASSIMILATION**

TIME-DIVERGENCE HYPOTHESIS [ecology]

Soulé, M. 1973. The epistasis cycle: a theory of marginal populations. *Annu. Rev. Ecol. Syst.* 4:165–87.

TIME LAW OF INTERSEXUALITY [genetics]

Goldschmidt, R. 1938. The time law of intersexuality. *Genetics* 20:1–59.

Allen, G. E. 1974. Opposition to the Mendelian-chromosome theory: the physiological and developmental genetics of Richard Goldschmidt. *J. Hist. Biol.* 7:49–92.

TIME-SPACE MODEL [species interactions]

Allen, J. C. 1975. Mathematical models of species interactions in time and space. *Amer. Nat.* 109:319–42.

Cf. HASSELL-VARLEY MODEL; LOTKA-VOLTERRA MODEL; NICHOLSON-BAILEY MODEL

TIME-STABILITY HYPOTHESIS [of community structure]

see **STABILITY-TIME HYPOTHESIS**

TIME THEORY [of species diversity]

Pianka, E. R. 1966. Latitudinal gradients in species diversity: a review of concepts. *Amer. Nat.* 100:33–46.

TOLERANCE, LAW OF [ecology; physiology]

Shelford, V. E. 1911. Physiological animal geography. *J. Morph.* 22:551–618.

______. 1913. The reactions of certain animals to gradients of evaporating power and air. A study in experimental ecology. *Biol. Bull.* 25:79–120.

Schafer, J. F. 1971. Tolerance to plant disease. *Annu. Rev. Phytopath.* 9:235–52.

Pianka, E. R. 1978. *Evolutionary Ecology*. New York: Harper & Row, chap. 1.

Valentine, D. H. 1978. Ecological criteria in plant taxonomy. *Essays in Plant Taxonomy*. Edited by H. E. Street. London & New York: Academic Press, pp. 1–18.

TOLERANCE, THEORY OF [plant geography]

Good, R. 1931. A theory of plant geography. *New Phytol.* 30:149–71.

Good, R. 1974. *The Geography of the Flowering Plants*. 4th ed. London: Longman, chap. 22.

TRACE NUTRIENT CONCEPT [nutrition]

Funk, C. 1912. The etiology of deficiency disease. *J. St. Med.* 20:341–65.

Ihde, A. J., & S. L. Becker. 1971. Conflict of concepts in early vitamin studies. *J. Hist. Biol.* 4:1–33.

TRANSFERENCE OF FUNCTION [evolution; morphology]

Corner, E. J. H. 1958. Transference of function. *J. Linn. Soc. Lond. Bot.* 56:33–40.

Stebbins, G. L. 1970. Transference of function as a factor in the evolution of seeds and their accessory structures. *Isr. J. Bot.* 19:59–70.

______. 1976. Seeds, seedlings, and the origin of angiosperms. *Origin and Early Evolution of Angiosperms.* Edited by C. B. Beck. New York: Columbia University Press, pp. 300–11.

TRANSFORMATION [genetics]

Avery, O. T., C. M. MacLeod, & M. McCarty. 1944. Studies on the chemical nature of the substance inducing transformation of pneumococcal types. *J. Exp. Med.* 79:137–57.

Portugal, F. H., & J. S. Cohen. 1977. *A Century of DNA.* Cambridge: M.I.T. Press, chap. 7.

TRANSFORMATION THEORY [morphogenesis]

Thompson, D'A. W. 1942. *On Growth and Form.* 2d ed. Cambridge: Cambridge University Press, chap. 18.

Woodger, J. H. 1945. On biological transformations. *Essays on Growth and Form Presented to D'Arcy Wentworth Thompson.* Edited by W. E. LeGros Clark & P. B. Medawar. Oxford: Clarendon Press, pp. 93–120.

Richards, O. W. 1955. D'Arcy W. Thompson's mathematical transformation and thc analysis of growth. *Ann. N.Y. Acad. Sci.* 63:456–73.

Rosen, R. 1967. *Optimality Principles in Biology.* London: Butterworths, p. 87.

Brooks, J. L. 1972. Extinction and the origin of organic diversity. *Trans. Conn. Acad. Arts Sci.* 44:17–56.

Dullemeijer, P. 1974. *Concepts and Approaches in Animal Morphology.* Assen, Netherlands: Van Gorcum, pp. 39, 196.

TRANSGENOSIS [genetic recombination]

Doy, C. H. 1972. Transfer and expression (transgenosis) of bacterial genes in plant-cells. *Search* 3:447–48.

______, P. M. Gresshoff, & B. G. Rolfe. 1973. Biological and molecular evidence for the transgenosis of genes from bacteria to plant cells. *Proc. Natl. Acad. Sci.* 70:723–26.

TRANSLATION ERROR-AMBIGUITY THEORY [of the origin of the genetic code] *see* **AMBIGUITY REDUCTION THEORY**

TRANSLOCON HYPOTHESIS [in immunoglobulin production]

Gally, J. A., & G. M. Edelman. 1972. The genetic control of immunoglobulin synthesis. *Annu. Rev. Genet.* 6:1–46.

TRANSMUTATION *see* **DARWIN'S THEORY OF TRANSMUTATION; GENE CONVERSION; ORGANIC CHANGE, THEORY OF**

TRANSPIRATIONAL PULL [plant physiology]
= transpiration

Grew, N. 1682. *The Anatomy of Plants.* London: Printed by W. Rawlins for the author. Vol. III.

Hales, S. 1727. *Vegetable Staticks: or, an Account of Some Statical Experiments on the Sap in Vegetables.* London: Innys.

Dixon, H. H. 1924. *The Transpiration Stream.* London: University of London Press.

Curtis, O. F. 1926. What is the significance of transpiration? *Science* 63:267–71.

Clements, H. F. 1934. Significance of transpiration. *Plant Physiol.* 9:165–72.

Sutcliffe, J. 1968. *Plants and Water.* New York: St. Martin's, chap. 6.

TRANSREPLICATION *see* **GENE CONVERSION**

TRAP THEORY [light sensitivity of cave animals]

Ludwig, W. 1942. Zur evolutorischen Erklärung der Höhlentiermerkmale durch Allelelimination. *Biol. Zblt.* 62:447–55.

Janzer, W., & W. Ludwig. 1952. Versuche zur evolutorischen Enstehung der Höhlentiermerkmale. *Z. indukt. Abstammungs Vererbungsl.* 84:462–79.

Barr, T. C., Jr. 1968. Cave ecology and the evolution of Troglobites. *Evol. Biol.* 2:35–102.

TRENDS *see* **ORTHOGENESIS**

TRIPARTITE COEVOLUTIONARY STRATEGY [disperser-pollinator-angiosperm association]

Regal, P. J. 1977. Ecology and evolution of flowering plant dominance. *Science* 196:622–29.

TRIPARTITE THEORY [origin of maize]

Mangelsdorf, P. C., & R. G. Reeves. 1939. The origin of Indian corn and its relatives. *Texas Agric. Exp. Sta. Bull.* 574:1–315.

Randolph, L. F. 1976. Contributions of wild relatives of maize to the evolutionary history of domesticated maize: a synthesis of divergent hypotheses I. *Econ. Bot.* 30:321–45.

TRITUBERCULAR THEORY *see* **COPE-OSBORN THEORY OF TRITUBERCULY**

TROPHIC-DYNAMIC CONCEPT (of energy flow in plant and animal communities)

Lindeman, R. L. 1942. The trophic dynamic concept of ecology. *Ecology* 23:399–418.

Odum, E. P. 1968. Energy flow in ecosystems: a historical review. *Amer. Zool.* 8:11–18.

Cook, R. E. 1977. Raymond Lindeman and the trophic-dynamic concept in ecology. *Science* 198:22–26.

Ulanowicz, R. E., & W. M. Kemp. 1979. Toward canonical trophic aggregations. *Amer. Nat.* 114:871–83.

Cf. ENERGY FLOW

TROPHIC LEVEL CONCEPT [ecology]

Lindeman, R. L. 1942. The trophic dynamic concept of ecology. *Ecology* 23:399–418.

Kozlovsky, D. G. 1968. A critical evaluation of the trophic level concept. I. Ecological efficiencies. *Ecology* 49:48–60.

Wiegart, R. G., & D. F. Owen. 1971. Trophic structure, available resources and population density in terrestrial vs. aquatic ecosystems. *J. Theor. Biol.* 30:69–81.

Rigler, F. H. 1975. The concept of energy flow and nutrient flow between trophic levels. *Unifying Concepts in Ecology.* Edited by W. H. Van Dobben & R. H. Lowe-McConnell. The Hague: Junk, pp. 15–26.

TROPISMS, THEORY OF [animal and plant orientation]

= Loeb's theory

Loeb, J. 1913. Die Tropismen. *Handb. Vergleich. Physiol.*, 4.

______. 1918. *Forced Movements, Tropisms, and Animal Conduct.* Philadelphia: Lippincott.

Weevers, T. 1949. *Fifty Years of Plant Physiology.* Amsterdam: Scheltema & Holkema; Waltham, Mass.: Chronica Botanica, pp. 231–60.

Fraenkel, G. S., & D. S. Gunn. 1961. *The Orientation of Animals.* Oxford: Clarendon Press.

Gussin, A. E. S. 1963. Jacques Loeb: the man and his tropism theory of animal conduct. *J. Hist. Med.* 18:321–36.

TROW'S EQUATION [chromosomal crossing-over]

Trow, A. H. 1913. Forms of reduplication: primary and

secondary. *J. Genet.* 2:313–24.

Kushev, V. V. 1974. *Mechanisms of Genetic Recombination.* New York: Consultants Bureau, p. 9.

TUBULAR CARPEL THEORY [of carpel origin] *see* **PELTATE CARPEL THEORY**

TUNICA-CORPUS CONCEPT [of apical organization in vascular plants]

Schmidt, A. 1924. Histologische Studien an phanerogamen Vegetationspunkten. *Bot. Arch.* 8:345–404.

Esau, K. 1977. *Anatomy of Seed Plants.* 2d ed. New York: Wiley, chap. 16.

TURING'S DIFFUSION REACTION THEORY [morphogenesis]

Turing, A. M. 1952. The chemical basis of morphogenesis. *Phil. Trans. Roy. Soc. B.* 237:37–72.

Wardlaw, C. W. 1953. A commentary on Turing's diffusion reaction theory of morphogenesis. *New Phytol.* 52:40–47.

______ . 1955. Evidence relating to the diffusion-reaction theory of morphogenesis. *New Phytol.* 54:39–48.

______ . 1968. *Essays on Form in Plants.* Manchester: Manchester University Press, chap. 10.

TURNOVER [ecology]

Leopold, A. C. 1975. Aging, senescence, and turnover in plants. *Bioscience* 25:659–62.

TURNOVER HYPOTHESIS [of island biogeography] *see* **ISLAND BIOGEOGRAPHY, THEORY OF**

TWO-CENTER EFFECT [in nerve growth]

Weiss, P. 1952. "Attraction fields" between growing tissue cultures. *Science* 115:293–95.

______ . 1955. Nervous system (neurogenesis). *Analysis of Development.* Edited by B. H. Willier, P. A. Weiss, & V. Hamburger, Philadelphia: Saunders pp. 346–401. (1971 reprint, New York: Hafner.)

Cf. ONE-CENTER EFFECT

TWO-FACTOR THEORY OF INHIBITION [behavior]

Thompson, M. E. 1960. A two-factor theory of inhibition. *Psychol. Rev.* 67:200–06.

Cf. MONOTONY PRINCIPLE

TWO-LOCUS HYPOTHESIS or THEORY [genetics]

Lewontin, R. C., & K. Kojima. 1960. The evolutionary dynamics of complex polymorphisms. *Evolution*. 14:458–72.

Bodmer, W. F., & P. Parsons. 1962. Linkage and recombination in evolution. *Adv. Genet.* 11:1–100.

Franklin, I., & R. C. Lewontin. 1970. Is the gene the unit of selection? *Genetics* 65:707–34.

Karlin, S., & D. Carmelli. 1975. Some population genetic models combining artificial and natural selection pressures. II. Two locus theory. *Theor. Pop. Biol.* 7:123–48.

Weir, B. S., & C. C. Cockerham. 1977. Two-locus theory in quantitative genetics. *Proc. Int. Conf. Quant. Genet.* Aug. 16–21, 1976. Edited by E. Pollack, O. Kempthorne, & T. B. Bailey. Ames: Iowa State University Press, pp. 247–69.

Cf. ADAPTIVE TOPOGRAPHY; GAMETIC DISEQUILIBRIUM; OUTLAW GENES

TWO PROTEINS HYPOTHESIS [genetics]

Bukhari, A. I. 1976. Bacteriophage MU as a transposition element. *Annu. Rev. Genet.* 10:389–412.

Cf. ONE PROTEIN HYPOTHESIS

TYPE [morphology; classification]

Goethe, J. W. von. 1790. *Versuch die Metamorphose der Pflanzen zu erklaren*. Gotha: Ettinger. Transl. 1946 by A. Arber in *Goethe's Botany*. Waltham, Mass.: Chronica Botanica, vol. 10, no. 2, pp. 63–126.

Zangerl, R. 1948. The methods of comparative anatomy and its contribution to the study of evolution. *Evolution* 2:351–74.

Arber, A. 1950. *The Natural Philosophy of Plant Form*. Cambridge: Cambridge University Press, pp. 59*ff*.

Mayr, E. 1957. Species concepts and definitions. *The Species Problem*. Edited by E. Mayr. Washington: American Association for the Advancement of Science, pp. 1–22.

Waddington, C. H. 1957. *The Strategy of the Genes*. London: Allen & Unwin, pp. 80*ff*.

Coleman, W. 1962. Georges Cuvier, biological variation and the fixity of species. *Arch. Int. Hist. Sci.* 15:315–31.

Farber, E. 1964. Theories of types in the history of sciences. *J. Wash. Acad. Sci.* 54:349–56.

Farber, P. L. 1976. The type-concept in zoology during the first half of the nineteenth century. *J. Hist. Biol.* 9:93–119.

TYPOLOGICAL SPECIES CONCEPT *see* **SPECIES CONCEPTS**

TYPOSTROPHY [evolution]

Riedl, R. 1977. A systems-analytical approach to macro-evolutionary phenomena. *Q. Rev. Biol.* 52:351–70.

Cf. HOLLOW CURVES

[U]

ULTIMATE AND PROXIMATE FACTORS [evolution]

Baker, J. R. 1938. The evolution of breeding seasons. *Evolution: Essays Presented to E. S. Goodrich.* Edited by G. R. de Beer. Oxford: Clarendon Press, pp. 161–77.

Immelmann, K. 1972. Erörterungen zur Definition und Anwendbarkeit der Begriffe "ultimate factor," "proximate factor," und "Zeitgeber." *Oecologia* 9:259–64.

UMBRELLA PRINCIPLE [physiology of disease resistance]

Horsfall, J. G., & A. E. Diamond. 1957. Interactions of tissue sugar, growth substances and disease susceptibility. Z. *Pflanzenkr. Pflanzenschutz* 64:415–21.

UNCERTAINTY PRINCIPLE [of Systematics]

Brown, W. L., Jr., & E. O. Wilson. 1956. Character displacement. *Syst. Zool.* 5:49–64.

Brown, W. L., Jr. 1957. Centrifugal speciation. *Q. Rev. Biol.* 32:247–77.

UNEQUAL RECOMBINATION HYPOTHESIS [genetics; evolution]

Smith, G. P. 1973. Unequal crossover and the evolution of multigene families. *Cold Spr. Harb. Symp. Quant. Biol.* 38:507–13.

Tartof, K. D. 1975. Redundant genes. *Annu. Rev. Genet.* 9:355–85.

UNIDIRECTIONAL PROGRESSION [primate evolution]

Minkoff, E. C. 1974. The direction of lower primate evolution: an old hypothesis revived. *Amer. Nat.* 108:519–32.

Cf. SMITH-JONES HYPOTHESIS

UNIFORMITARIANISM [evolution]

Hooykaas, R. 1957. The parallel between the history of the earth and the history of the animal world. *Arch. Int. Hist. Sci.* 38:3–18.

Cannon, W. F. 1960. The uniformitarian-catastrophist debate. *Isis* 51:38–55.
Wilson, L. G. 1964. The development of the concept of uniformitarianism in the mind of Charles Lyell. *Proc. 10th Int. Congr. Hist. Sci.* 2:993–96.
Simpson, G. G. 1970. Uniformitarianism. An inquiry into principle, theory and method in geohistory and biohistory. *Essays in Evolution and Genetics in Honor of Theodosius Dobzhansky*. Edited by M. K. Hecht & W. C. Steere. (Suppl. to *Evolutionary Biology*.) Pp. 43–96.

Cf. CATASTROPHISM

UNIQUELY DERIVED CHARACTER CONCEPT [derivation of cladistic relationships]

LeQuesne, W. J. 1969. A method of selection of characters in numerical taxonomy. *Syst. Zool.* 18:201–05.
______. 1972. Further studies on the uniquely derived character concept. *Syst. Zool.* 21:281-88.
Colless, D. H. 1973. A note on LeQuesne's uniquely derived character concept. *Syst. Zool.* 22:320–21.

UNIQUELY EVOLVED CHARACTER CONCEPT [systematics; evolution]

LeQuesne, W. J. 1974. The uniquely evolved character concept and its cladistic application. *Syst. Zool.* 23:513–17.
______. 1977. The uniquely evolved character concept. *Syst. Zool.* 26:218–20.
Farris, J. S. 1977. Some further comments on LeQuesne's methods. *Syst. Zool.* 26:220–23.

Cf. DOLLO'S LAW

UNIT-MEMBRANE THEORY [biophysics]

Robertson, J. D. 1959. The ultrastructure of the cell membranes and their derivatives. *Biochem. Soc. Symp.* 16:3–43.
______. 1960. The molecular structure and contact relationships of cell membranes. *Prog. Biophys. Biophys. Chem.* 10:343–418.
______. 1964. Unit membranes: a review with recent new studies of experimental alterations and a new subunit structure in synaptic membranes. *Cellular Membranes in Development*. Edited by M. Locke. New York: Academic Press, pp. 1–81.
______. 1967. Origin of the unit membrane hypothesis. *Protoplasma* 63:218–45.
Korn, E. D. 1966. Structure of biological membranes. *Science* 153:1491–98.

UP-GRADE THEORY [of bryophyte evolution]
Richards, P. W. 1959. Bryophyta. *Vistas in Botany*. Edited by W. B. Turrill. New York: Pergamon, pp. 387–420.

UPLAND THEORY [of angiosperm origin]
Axelrod, D. I. 1960. The evolution of flowering plants. *Evolution after Darwin*. vol. I: *The Evolution of Life*. Edited by S. Tax. Chicago: University of Chicago Press, pp. 227–305.

[V]

VAN'T HOFF RULE [relating biological processes to temperature]
Van't Hoff, T. H. 1884. *Études de Dynamic Chimique*. Amsterdam: Müller.
Allee, W. C., A. E. Emerson, O. Park, T. Park, & K. P. Schmidt. 1949. *Principles of Animal Ecology*. Philadelphia: Saunders, p. 107.
Cf. ARRHENIUS' FORMULA

VAVILOV'S LAW *see* **HOMOLOGOUS SERIES, LAW OF**

VERNALIZATION [physiology]
Lysenko, T. D. 1928. Effect of the thermal factor on length of phases in development of plants. (In Russian.) *Trudy Azerbajdžansk. Opytn. Sta.* no. 3.
Whyte, R. O. 1948. History of research in vernalization. *Vernalization and Photoperiodism: A Symposium*. Edited by A. E. Murneek & R. O. Whyte. Waltham, Mass.: Chronica Botanica, pp. 1–38.

VERSCHMELZUNG THEORY *see* **FUSION THEORY** [of heredity]

VERTEBRAE, LAW OF *see* **JORDAN'S RULE**

VERTEBRAL THEORY [of skull origin] *see* **GOETHE'S VERTEBRAL THEORY**

VICARIANCE THEORY [biogeography]

Croizat, L. 1958. *Panbiogeography.* 2 vols. Caracas: By the Author.

Rosen, D. E. 1975. A vicariance model of Caribbean biogeography. *Syst. Zool.* 24:431–64.

VICARIOUS SPECIES [biogeography]

= species pairs = vicariism

Cain, S. A. 1944. *Foundations of Plant Geography.* New York: Harper, chap. 18.

Good, R. 1974. *The Geography of the Flowering Plants.* 4th ed. London: Longmans, pp. 250–53.

VIRUS

Hughes, S. S. 1977. *The Virus: A History of the Concept.* New York: Science History Publications.

VIRUS THEORY [of cancer]

Rous, P. 1910. A transmissible avian neoplasm. (Sarcoma of the common fowl.) *J. Exp. Med.* 12:696–705.

Needham, J. 1942. *Biochemistry and Morphogenesis.* Cambridge: Cambridge University Press, pp. 267*ff*.

Rather, L. J. 1978. *The Genesis of Cancer: A Study in the History of Ideas.* Baltimore: Johns Hopkins Press.

Cf. ONCOGENE HYPOTHESIS

VOCABULARY EXPANSION THEORY [of the origin of the genetic code]

Crick, F. H. C. 1967. Origin of the genetic code. *Nature* 213:119.

Jukes, T. H. 1967. Indications of an evolutionary pathway in the amino acid code. *Biochem. Biophys. Res. Comm.* 27:573–78.

Woese, C. 1969. Models for the evolution of codon assignments. *J. Mol. Biol.* 43:235–40.

Fitch, W. M. 1973. Aspects of molecular evolution. *Annu. Rev. Genet.* 7:343–80.

VOLTERRA-GAUSE PRINCIPLE *see* **COMPETITIVE EXCLUSION**

VON BAER'S LAWS [embryonic resemblance in animals]

Von Baer, K. E. 1828–37. *Uber Entwickslungsgeschichte der Thiere. Beobachtung und Reflexion.* 2 vols. Königsberg: Gebrüder Borntraeger, pt. 1.

———. 1853. On the development of animals, with observations and reflections. *Scientific Memoirs*, ser. 2. Edited by A. Henfry & T. H. Huxley. London: Taylor & Francis, pp. 186–238.

———. 1866. De la découverte de larves qui se propagent. *Bull. Acad. Imp. Sci. St. Petersbourg* 9:64–137.

Darwin, C. 1872. *The Origin of Species*. 6th ed. London: Murray, p. 390.

Sedgwick, A. 1894. On the law of development commonly known as von Baer's law, and on the significance of ancestral rudiments in embryonic development. *Q. J. Microsc. Soc.* 36:35–52.

de Beer, G. R. 1958. *Embryos and Ancestors*. 3d ed. Oxford: Clarendon Press, chap. 1.

Thompson, D'A. W. 1942. *On Growth and Form*. 2d ed. Cambridge: Cambridge University Press, pp. 86*ff*.

Oppenheimer, J. 1967. *Essays in the History of Embryology and Biology*. Cambridge: M.I.T. Press.

Ballard, W. W. 1976. Problems of gastrulation: real and verbal. *Bioscience* 26:36–39.

Ospovat, D. 1976. The influence of Karl Ernst von Baer's embryology, 1828–1859; a reappraisal in light of Richard Owen's and William B. Carpenter's "Paleontological Application of 'Von Baer's Law.'" *J. Hist. Biol.* 9:1–28.

Gould, S. J. 1979. On the importance of heterochrony for evolutionary biology. *Syst. Zool.* 28:224–26.

Cf. BIOGENESIS

[W]

WAGNER DIVERGENCE INDEX [systematics]

= Wagner trees = Wagner network

Hardin, J. W. 1957. A revision of the American Hippocastanaceae. *Brittonia* 9:145–71.

Wagner, W. H. 1961. Problems in the classification of ferns. *Recent Advances in Botany*. Toronto: University of Toronto Press, 1 (sect. 9): 841–44.

Scora, R. W. 1966. The evolution of the genus *Monarda* (Labiatae). *Evolution* 20:185–90.

———. 1967. Divergence in *Monarda* (Labiatae). *Taxon* 16:449–505.

Farris, J. S. 1970. Methods for computing Wagner trees. *Syst. Zool.* 19:83–92.

Kluge, A. G., & J. S. Farris. 1969. Quantitative phyletics and

the evolution of annurans. *Syst. Zool.* 18:1–32.
Lundberg, J. G. 1972. Wagner networks and ancestors. *Syst. Zool.* 21:398–413.
Cf. PHYLOGENETIC TREE

WAGNER'S LAW *see* **JORDAN'S LAW** [of geographic speciation]

WAHLUND EFFECT or PRINCIPLE [mimicry of inbreeding]
Wahlund, S. 1928. Zusammenzetzung von Population und Korrelationserscheinungen von Standpunkt der Vererbungslehre ans betrachtet. *Hereditas* 11:65–106.
Sinnock, P. 1975. The Wahlund effect for the two-locus model. *Amer. Nat.* 109:565–70.
Spiess, E. B. 1977. *Genes in Populations*. New York: Wiley, chap. 12.

WALLACE'S LAW [evolution]
Wallace, A. R. 1855. On the law which has regulated the introduction of new species. *Ann. Mag. Nat. Hist.* 16:184–96.
Beddall, B. G. 1972. Wallace, Darwin, and Edward Blyth: further notes on the development of evolution theory. *J. Hist. Biol.* 5:153–58.

WALLACE'S LINE [biogeography]
Wallace, A. R. 1890. *The Malay Archipelago*. 10th ed. London: Macmillan, pp. 8–9. (1962 reprint; New York: Dover.)
Schuster, R. M. 1972. Continental movements, "Wallace's Line" and Indmalayan-Australasian dispersal of land plants: some eclectic concepts. *Bot. Rev.* 38:3–86.
———. 1976. Plate tectonics and its bearing on the geographical origin and dispersal of angiosperms. *Origin and Early Evolution of Angiosperms*. Edited by C. B. Beck. New York: Columbia University Press, pp. 48–138.

WALLACE'S RULES [of mimicry]
Folsom, J. W. 1926. *Entomology with Reference to Its Biological and Economic Aspects*. 3d ed. Philadelphia: Blakiston, p. 203.
Goldschmidt, R. B. 1945. Mimetic polymorphism. A controversial chapter of Darwinism. *Q. Rev. Biol.* 20:147-64, 205–30.
Allee, W. C., A. E. Emerson, O. Park, T. Park, & K. P. Schmidt. 1949. *Principles of Animal Ecology*. Philadelphia: Saunders, p. 670.
Cf. MIMICRY

WARBURG'S MEMBRANE THEORY OF RESPIRATION

Warburg, O. 1910. Über die Oxydationen in lebenden Zellen nach Versuchen am Seeigelei. *Hoppe-Seyl. Z.* 66:305–40.

Kohler, R. E. 1973. The background to Otto Warburg's conception of the *Atmungsferment*. *J. Hist. Biol.* 6:171–92.

Cf. ATMUNGSFERMENT; NUCLEAR THEORY OF RESPIRATION

WATSON-CRICK HYPOTHESIS [genetic code; structure of DNA]

Watson, J. D., & F. C. H. Crick. 1953. A structure for deoxyribose nucleic acid. *Nature* 171:737–38.

______. 1953. The structure of DNA. *Cold Spr. Harb. Symp. Quant. Biol.* 18:123–31.

______. 1953. Genetical implications of the structure of deoxyribonucleic acid. *Nature* 171:964–67.

Watson, J. D. 1968. *The Double Helix*. New York: Antheneum.

Commoner, B. 1968. Failure of the Watson-Crick theory as a chemical explanation of inheritance. *Science* 220:334–40.

Olby, R. 1974. *The Path to the Double Helix*. New York: Collier-Macmillan.

Portugal, F. H., & J. S. Cohen. 1977. *A Century of DNA*. Cambridge: M.I.T. Press

WATSON'S RULE *see* **MOSAIC EVOLUTION**

WAVE THEORY [of bison evolution]

Skinner, M. F., & O. C. Kaiser. 1947. The fossil bison of Alaska and preliminary revision of the genus. *Bull. Amer. Mus. Nat. Hist.* 89:123–256.

Guthrie, R. D. 1970. Bison evolution and zoogeography in North America during the Pleistocene. *Q. Rev. Biol.* 45:1–15.

WAVES OF CONTRACTION *see* **CONTRACTION WAVE CONCEPT**

WAVES OF LIFE *see* **POPULATION WAVES**

WEBER'S LAW [of muscle action]

Weber, E. F. 1846. Muskelbewegung. *Rudolph Wagner Handwörterbuch*. Braunschweig: Vieweg.

Fick, R. 1911. *Handbuch der Anatomie und Mechanik der Gelenke unter Berücksichtigung der bewegenden Muckeln*. Jena: Fischer.

Dullemeijer, P. 1974. *Concepts and Approaches in Animal Morphology*. Assen, Netherlands: Van Gorcum, p. 85.

WILLIS'S THEORY *see* **AGE AND AREA**

WING RULE [wing shape partially dependent on climate] = RENSCH'S RULE

Kipp, F. A. 1942. Über Flügelbau und Wanderzug der Vögel. *Biol. Zblt.* 62:289–99.

Mayr, E. 1963. *Animal Species and Evolution.* Cambridge: Belknap Press of Harvard University Press, chap. 11.

WISDOM OF THE BODY *see* **PHYSIOLOGICAL HOMEOSTASIS**

WOBBLE HYPOTHESIS [genetic code]

Crick, F. H. C. 1966. Codon-anticodon pairing: the wobble hypothesis. *J. Molec. Biol.* 19:548–55.

Caskey, C. T. 1970. The universal RNA genetic code. *Q. Rev. Biophys.* 3:295–326.

WOLF-IN-SHEEP'S-CLOTHING STRATEGY [ecology]

Eisner, T., K. Hicks, & M. Eisner. 1978. "Wolf-in-sheep's-clothing" strategy of a predaceous insect larva. *Science* 199:790–94.

WOLFF'S LAW [bone structure modification]

Wolff, J. 1870. Uber die innere Architektur der knochen und ihre Bedeutung für die frage vom Knochenwachstum. *Virchows Arch. Path. Anat.* 50:389–450.

———. 1892. *Das Gesetz der Transformation der Knochen.* Berlin: Hirschwald.

Enlow, D. H. 1968. Wolff's law and the factor of architectonic circumstance. *Amer. J. Orthod.* 54:803–22.

Chamay, A., & P. Tschantz. 1972. Mechanical influences in bone remodeling. Experimental research on Wolff's law. *J. Biomech.* 5:173–80.

Pugh, J. W., R. M. Rose, & E. L. Rodin. 1973. A possible mechanism of Wolff's law: trabecular microfractures. *Arch. Int. Physiol. Biochim.* 81:27–40.

WOLFF'S THEORY *see* **EPIGENESIS**

WRIGHT'S ADAPTIVE SURFACE *see* **ADAPTIVE SURFACE**

WRIGHT'S PRINCIPLE OF MAXIMIZATION OF W *see* **ADAPTIVE PEAK; FITNESS** [Darwinian]

[Y]

YOU-FIRST PRINCIPLE [behavior]

Hamilton, W. D. 1971. Geometry for the selfish herd. *J. Theor. Biol.* 31:295–311.

Brown, J. L. 1975. *The Evolution of Behavior*. New York: Norton, p. 144.

[Z]

ZAHNREIHE THEORY [of tooth replacement]

Edmund, A. G. 1960. Tooth replacement phenomena in the lower vertebrates. *Contributions of the Life Sciences Division, Royal Ontario Museum* 52:1–190.

———. 1962. Sequence and rate of tooth replacement in the Crocodilia. *Contributions of the Life Sciences Division, Royal Ontario Museum* 56:1–42.

Osborn, J. W. 1970. New approach to Zahnreihen. *Nature* 225:343–46.

———. 1974. On tooth succession in *Diademodon*. *Evolution* 28:141–57.

DeMar, R. 1972. Evolutionary implications of *Zahnreihen*. *Evolution* 26:435–50.

ZEEMAN'S CATASTROPHE MACHINE [biodynamics]

Zeeman, E. C. 1972. A catastrophe machine. *Towards a Theoretical Biology*. Edited by C. H. Waddington. Edinburgh: Edinburgh University Press, 4:276–82.

Poston, T., & A. E. R. Woodcock. 1973. Zeeman's catastrophe machine. *Proc. Camb. Philos. Soc.* 74:217–26.

Stewart, I. 1975. The seven elementary catastrophes. *New Sci.* 68:447–54.

Cf. CATASTROPHE THEORY; THOM'S THEORY OF CATASTROPHES

ZONES OF INTERGRADATION *see* **INTERGRADATION ZONES**

ZOO HYPOTHESIS [origin of life on earth]

Ball, J. A. 1973. The zoo hypothesis. *Icarus* 19:347–49.

ZOOGEOGRAPHIC SUBREGIONS [biogeography]

Horton, D. 1973. The concept of zoogeographic subregions. *Syst. Zool.* 22:191–95.

ZYMASE HYPOTHESIS [supporting the enzyme theory of life]

Buchner, H. 1897. Die Bedeutung der activen löslichen Zellprodukte für den Chemismus der Zelle. *Münch. Med. Wochens.* 44:299–302, 321–22.

______ . 1897. Alkoholische Gährung ohne Hefezellen. *Ber. Deut. Chem. Ges.* 30:117–24.

Kohler, R. E. 1972. The reception of Eduard Buchner's discovery of cell-free fermentation. *J. Hist. Biol.* 5:327–53.

Cf. ENZYME THEORY; PROTOPLASM THEORY